TEAM
BUILDING

TEAM BUILDING

A Memoir about Family and
the Fight for Workers' Rights

BEN GWIN

Belt Publishing

Copyright © 2023 by Ben Gwin

Printed in the United States of America
First edition 2023
1 2 3 4 5 6 7 8 9

ISBN: 9781953368331

Belt Publishing
13443 Detroit Avenue, Lakewood, OH 44107
www.beltpublishing.com

Cover art by David Wilson
Book design by Meredith Pangrace

For my father

TABLE OF CONTENTS

PART I

March 2019
SHORT-TERM DISABILITY

I left the hospital with a catheter and a bottle of Oxycodone. I put my hand down the front of my sweatpants and held the empty urinary bag next to my thigh so it wouldn't jostle while I walked. When I was far enough from the exit, I fished a half-smoked cigarette from my pack and sucked it down while I waited for my mom to pull the car around. It tasted terrible.

A day earlier, I had checked into the hospital for hemorrhoid surgery, but complications from the anesthesia left me unable to pee. They kept me overnight, waiting to see if I could use the bathroom on my own. One nurse gave me a plastic jug to keep next to the bed. She said I should have been able to fill the jug about halfway, but when I tried, I failed. Three times nurses used a catheter to empty my bladder. After the fourth, they left the catheter in. It hurt. In six more days, if there were no further problems, they said they'd remove it.

I'd scheduled my procedure for the Thursday before the opening weekend of the NCAA basketball tournament. I was supposed to be back at work on Monday. I texted my manager, Donald, to let him know I'd be out longer than I'd expected:

Not sure when I'll be back, I wrote.

He replied, *I will email directions for short-term disability. Hope you feel better.*

I ambled past a bulldozer surrounded by upturned earth, sitting empty in an adjacent lot. I leaned on a bus shelter, high on pain meds, and stared blankly at the clouds above the hill across the street. It smelled like rainwater and dirt—one of the first nice days of spring. I felt myself piss. The bag warmed against my leg.

When my mom pulled up, I eased into the passenger seat, and we headed to my apartment in Bloomfield, about three miles northeast of downtown Pittsburgh. I was lucky she could visit and help watch my daughter, Gracie, while I recovered. Gracie was having a tough time in middle school. I was having a tough time at work. It was a tough time all around.

"Never thought you'd wind up in Depends before your father," Mom said.

"Thanks, Mom."

I leaned my head against the window as we wove through North Oakland toward Liberty Avenue. We hit a pothole. I bounced on the seat, and the catheter shifted.

"Sorry," Mom said.

"It's fine."

Aside from some Vicodin after a root canal, I hadn't taken anything stronger than Tylenol since 2005. I'd been sober almost fourteen years. Outside, the city looked like it had soft edges.

The overnight stay at the hospital cost me $2,000, and I'd already lost out on three days' pay due to my job's time-off policy.

I was working as a data analyst at Google's Pittsburgh office in East Liberty, but I was employed through a third-party contractor called HCL. My position didn't include dedicated sick days, and I'd already used up all my PTO on job interviews that hadn't panned out. According to our operations manager, Nancy, the company didn't allow us to take single days of unpaid leave. If we ran out of PTO and needed additional time off, we had to take a whole week. Nancy said this had something to do with the payroll system.

Nothing at work really made sense to me. The policies seemed convoluted, unnecessarily restrictive, and unevenly applied. Instead of working Monday through Wednesday and then going in on Thursday for my surgery, I had to take the whole week off, unpaid. And I wouldn't be eligible for short-term disability until seven days after surgery, at which time I would have to send documentation proving I couldn't work. If I missed any time after those initial seven days, I would receive 70 percent of my regular pay. At my previous job, I would have been able to work from home after I had healed but before I was comfortable enough to go back to the office. But at HCL, only select team leads and upper management were allowed to work remotely. I was just an analyst. At least the pay periods had lined up; I would have two partial checks instead of missing an entire one. I could take the financial hit, but if another unexpected expense came up in the next few months, I'd be in trouble.

Back at my apartment, I emptied the piss bag into the toilet and got situated on the couch. I taped the tube sticking out of my urethra to my thigh, just below the bottom of the adult diaper, then looped it up and over the elastic of my sweatpants in a plastic S shape. There was a clip at the end that connected to a plastic bag I set in a bucket on the floor.

Every movement hurt. I waited until I couldn't stand the pain anymore, then took half a pill. I decided I would try to hold off during the day and only take them at night so I could sleep. More than anything, I was thankful I didn't have to go to work.

June 2018
FIRST DAY OF WORK

Nine months before my surgery, I drove Gracie over to the Boys and Girls Club in Lawrenceville around 7:30 a.m. and dropped her off.

"I start my new job today, honey." I told her. "Hopefully we can go on vacation more now. I'll pick you up at five for softball."

"OK," she said. "I'm glad you're happy, Dad."

"I love you," I said, and she was out the door with a wave.

I watched her walk down the street, up the steps, and inside the building. Then I headed down Butler toward Stanton Avenue. Traffic wasn't too bad when school was out for the summer. Lawrenceville was full of glossy new apartment buildings, many of which sat empty. Construction permits hung indifferently in the windows of row houses with worn brick facades. Many of the older buildings were being torn down to make room for new construction—in a suddenly chic neighborhood like this one, the lots were worth more than the buildings sitting on them.

Gracie was about to turn eleven, and during the summer months, she had softball three nights a week. Her home games were at Mellon Park, right across the street from my new office. She was nervous but excited about starting junior high at a new school in the fall. I thought about how terrible my junior high years had been, for both me and my parents. My friends and I stole liquor from our parents' cabinets and snuck out at night to smoke in the woods. My grades were bad, and I didn't listen. The summer before ninth grade, I went on Ritalin and things improved, though that presented its own set of problems.

Now, I was doing my best to provide Gracie with stability and consistency while Jane, her mom, tried to get her own life together. Ten years of Jane's heroin addiction, family court, and supervised visitations had worn me out, but things were starting to look better. Over the past year, Jane had been to jail, rehab, inpatient, and outpatient. But recently, she had passed every piss and hair test. And at our last hearing, the judge ordered that Gracie was to go up to Zelienople, a town thirty miles north of Pittsburgh, to see Jane every other weekend, unsupervised.

Just three years earlier, Jane had relapsed. She'd only recently been allowed to be alone with Gracie again. But now she had a good job waiting tables. And she was on state insurance that also covered Gracie. Between the insurance and child support, I was saving around $700 a month in expenses. I needed the help. The health insurance was especially important. Gracie had a ventricular septal defect and asthma. Jane's state plan covered her heart doctor appointments and her prescriptions.

During my previous job at Thermo Fisher Scientific, it always felt like I was a whisper away from being laid off. I worked on six-month contracts, and every time one contract was running down and I hadn't received my next one, I held my breath and started job searching until they finally renewed my position. Thermo Fisher was tough to leave because it paid well and I had time to write. Over the course of five years, I'd worked there as a contractor two different times, and I got along well with everyone. But in between those two stints, I had been laid off. When my unemployment ran out, I took a series of bad, low-paying, mostly part-time jobs until they rehired me.

I didn't want to go through that again at thirty-eight. So in my last flurry of job applications, I had applied with HCL to work on-site at Google. The job description was similar to what I was already doing, but Google also had a reputation as a great place to work.

Even though I wasn't technically a Google employee, I wasn't exactly a contractor either. The company considered me a "vendor," part of Google's extended workforce that fell under the Temp, Vendor, Contractor (TVC) umbrella. This arrangement meant Google didn't pay for my benefits and wasn't responsible for my working conditions; HCL was. It also meant Google was not subject to joint employment laws. Those regulations are intended to protect workers by preventing companies from hiring temporary workers to do permanent, full-time jobs without paying them the wages of full-time workers, providing health insurance, or offering other workplace protections. On paper, Google didn't manage vendors like me or compensate us in any way (aside from food, apparently) because that would mean they would have to give us the same benefits they gave their full-time employees. Google workers weren't even supposed to give us T-shirts or invite us to on-site, company happy hours.

One well-known case of a company violating joint employment laws happened in the late nineties at Microsoft. All the contractors at the company had worked for years without the benefits Microsoft offered to its own workers. They eventually formed an independent union and won a lawsuit, but most of them lost their jobs in the process. Their union was unaffiliated because they couldn't find an international union to back them, which was due in part to holes in US labor law that make it much harder for temps and contractors to unionize and, therefore, less likely for a union to put resources into organizing them. Microsoft eventually settled in a huge class action lawsuit.

When I took the job with HCL, I only had a basic understanding of labor law, and this stuff wasn't at the top of my mind. But it was hard to find a job with a principal employer and not through a recruiting firm. After the recession, more and more businesses started using temp agencies. Over half of Google's workforce is made up of TVCs. It's a model that fractures employees, keeps temporary workers in a precarious position, and keeps wages and insurance costs down. All these things make it harder to organize a union. Some agencies even just find another firm to hire the workers and do all the administrative work. They just take a cut of the money.

HCL's recruiter, Stephen, had told me they started people off at a relatively low salary—$40,000 a year—but that they then gave big raises once new hires proved themselves. I was told I could make between $35 and $50 an hour at Google and that I would get back to my previous salary of $52,000 in a couple years. I figured that if Google was a better company to work for than Thermo Fisher, the initial pay cut would outweigh the stress that came from working from one six-month contract to the next and having to rely on the Affordable Care Act for health insurance.

I took the recruiter and the interviewing manager at face value. I felt like I was making an informed decision.

———

Stanton Avenue wound through Highland Park to East Liberty Boulevard. I turned down Hamilton Avenue, parked a couple blocks away from the Google building, and walked the rest of the way. I got there in time for a cigarette. The offices were housed in a refurbished Nabisco factory in Bakery Square, a neighborhood that hadn't existed prior to 2010. Near the benches in the smoking area, there was a sign bearing a brief history of the development—uninspired marketing copy that tried to invoke a sense of progress and revitalization in a section of town that had been redeveloped to attract incoming wealthy, mostly white tech workers. Bakery Square was part of a bigger initiative to rebrand Pittsburgh as a new tech hub. Along with Google, there were offices in the city for Amazon, Microsoft, Uber, Apple, and dozens of robotics and autonomous vehicle startups. They offered some high-paying jobs, but they also cultivated an underclass of tech workers like those of us at HCL. The companies received tax incentives for putting down roots in Pittsburgh, but a lot of their workers were also starting to get priced out of what had been a cheap city with a decent arts scene and a relatively high quality of living. In the case of Bakery Square, Walnut Capital had built a glorified shopping mall surrounded by expensive apartments. This led to rising rents and the displacement of longtime residents of East Liberty, a historically Black neighborhood that had already been redlined and carved up by urban renewal.

Surrounding the old Nabisco factory was a plaza with a slew of chain stores, a bank, an LA Fitness, and a Spring Hill Suites. Even during the summer, a decidedly pleasant time of year in Pittsburgh, the place made me uneasy. Before working there, I only ever visited Bakery Square for my men's league baseball meetings, which were held in a nearby bar, or when Gracie had practice or games at the field at Mellon Park across the street. On the other side of Penn Avenue, there was a new gated community behind the second Google building. Development in Pittsburgh has never been equitable, and Walnut Capital certainly wasn't about to course correct now. I wondered if it was better to have a Bakery Square instead of empty buildings and vacant lots, and why those seemed to be the only options.

After my cigarette, I took the elevator up and sat down in the seventh-floor lobby. I introduced myself to Sam and Bobby, two other new hires who were starting that morning too. In true Pittsburgh fashion, I'd met Bobby at a literary event in Greensburg a couple years earlier. It was nice to run into someone who knew me as a writer. Bobby built websites and had worked on the school paper in college. Sam had been in international development and knew multiple languages. They were both about ten years younger than me and very personable. I felt awkward and old.

Sam said, "What is up with the website?"

HCL's website, which we'd all used for the onboarding process, was a glitchy nightmare. It had taken hours—unpaid— to get our background checks and information into the system.

"It's bad," I said. I picked up a Rubik's Cube from the table in front of me and clicked it around without paying attention to the colors.

"I could make a better one," Bobby said.

"Are you two starting at $40,000?" I asked. I always volunteered my salary information to coworkers. After being lowballed at previous jobs, I found it best to get it out in the open.

"I tried to get more, but they said no," Sam said.

"Same," said Bobby.

"Me too."

"Free lunch at least," said Sam.

"Plus snacks," Bobby said.

Our deputy general manager, Jeffrey, who spent most of his time working from home or out of HCL's Minnesota office, had emailed us after we'd been hired, hyping the availability of snacks at the job. His email said we didn't need to bring anything with us to work because everything we needed was already at the office.

"I hope the raises are good," I said.

I set the Rubik's Cube on the table and reached for my beat-up messenger bag. The zipper was broken, and the clasps didn't work, but it had been a Christmas gift from my parents, and I didn't want to get rid of it. I wasn't sure I would even need it today. But in case I had some downtime, I brought a notebook, the novel I was reading, and a folder full of story drafts I was working on.

Nancy, the operations manager and team lead of Catalogs, the team we were all going to join, came in through the game room.

"How is everyone?" she asked.

Everyone answered with some version of "great." She smiled and ushered us inside.

"If anyone wants to grab a water or a coffee, go ahead," she said.

No one did.

Nancy had interviewed each of us, and she had told us all the same things: her team was growing, there was room to advance, and it was possible that we could be converted to full-time Google employees. We were all told we'd get more paid time off and a higher salary if we put our time in and worked hard. The interviews had been in the offices adjacent to the game room, so on our way in, all of us had walked past a pool table, arcade games, and pinball machines. I wasn't good at pinball, but I loved it, and they had a *Terminator* machine.

Now, we walked through the game room and turned left. I followed Nancy, Sam, and Bobby down a staircase that was designed to look like a roller coaster to our pod in the open office. I had never worked in an open office before, but from the very beginning, I didn't like it. This particular one felt like something out of a George Saunders story. On our floor, all the conference rooms and offices were named after rides and attractions from Kennywood, a local amusement park. I'd never been there, but I knew the names of the rides from the commercials—there was Racer, Jack Rabbit, Log Jammer, and Exterminator. Nancy, like a proud parent telling us about her kid, pointed out the various amenities and led us to our desks. There were microkitchens (the nearest one was named "Potato Patch" after the Kennywood fry stand) filled with snacks and bottled beverages on both sides of the pod, she noted, pointing in the direction of one and then the other. I could tell she was a good saleswoman.

Our pod was made up of six desks in a horseshoe shape, all facing inward. Mine was on the end. Next to it was a small filing cabinet with a padded top, and stuck to the cabinet was a magnet the size of an index card with my name on it. The cabinet had a couple drawers and rings for file folders. A whiteboard on wheels served as a small barrier between the walkway and the empty desk across from mine, as if the previous occupant had been trying to build themselves a cubicle.

The air conditioner was on a little too high, and I felt cold and out of place. I set my bag under my desk.

All the desks had big monitors and new Chromebooks waiting for us, but they would be switched out for Macs by the end of the week because the Google computers didn't have enough processing power to run the Google tools effectively. Tech Stop, the Google IT department, was down the hall if we needed any other equipment. It consisted of a big counter, where the IT guy sat, and several bins of cords, wires, and adapters we could take if we needed them. They stopped handing out iPhone chargers for free after my first few months, presumably because too many folks were leaving with them.

I looked for the least conspicuous exits so, down the road, I could sneak out early if necessary. With the open floor plan, it would be hard to avoid bosses going in or out. It felt like a panopticon.

We set up our workstations and then went to orientation. Cynthia, a new hire on the Clustering team joined us. At orientation, we learned about our health insurance plan—a pretty good one by American standards. We were also told not to talk to full-time Google employees about anything for any reason, ever. When Nancy got to this point in the orientation, her tone grew more serious as she rattled off a list of rules:

Don't tell anyone you work for Google. You work for HCL.

Don't put Google on your LinkedIn or on social media. Say "HCL onsite at Google."

You can put Senior Analyst on your résumé. After a year, you can write Lead Analyst.

Don't stay for dinner too often, and if you do, don't eat too much. Don't take food home from the cafeteria.

Don't ask your Google program manager for anything. Don't ever talk to your Google supervisors. Don't talk to Google employees.

Don't ask your coworkers about how much money they make. (It is illegal to forbid salary discussions in the workplace, but we eventually found that this rule against discussing salary was listed in our elusive online employee handbook. I didn't find out about it for months, before the line mysteriously vanished.)

Don't sit in the same place for too long—the Googlers might want to sit there, and we don't want to take up too much space. Try to stay at your desk for most of the day, but feel free to walk around with your laptop to one of the lounges or other workstations.

Aside from four hours of annual online training, you can't work from home under any circumstances. We do offer flextime. You can't work four tens, but you can work four nines and a four. I really love the flexibility. You'll see everyone leaving around noon on Fridays. It's great.

Don't come into work before 7:00 a.m.

Don't stay in the office later than 7:00 p.m.

There is no overtime under any circumstances.

Don't walk too fast past Googlers.

Don't talk too loud near Googlers.

If you have a problem with any full-time Google employees, the FTEs, go to your manager and they will escalate the issue to Google HR if necessary. Don't go to Google yourself.

You have three free hours of parking in the garage, but you are not allowed to leave and come back continuously throughout the day. Don't take up the FTEs' parking spaces.

We have ten paid holidays. There are two unpaid holidays when the office is closed, and we don't get paid time off: those days are MLK Day and Presidents' Day. On those two weeks, everyone works thirty-six hours in four days and uses four hours of PTO. Or you can use a full day of PTO and work thirty-two hours. It's not so bad. There's nothing we can do about these rules. Google enforces them.

We have a chat for funny memes. Feel free to share memes. It's fun!

We all sat quietly, still half-enthralled by the decadent office, the average health insurance, and the above average free food. At Thermo Fisher, I had kept a loaf of bread and a jar of peanut butter in my cubicle for lunches, along with a box of Clif Bars for breakfast. Burnt breakroom coffee did the job. Every Thursday, I'd get a salad from the cafeteria and a Starbucks coffee with the other contractors. We'd walk around outside, smoke cigarettes, and complain about our boss. We would share notes on our individual progress, and we always made sure to hand in our weekly quota of spreadsheets at the same time so that no one was going too fast. We needed the work to last as long as possible while still hitting our deadlines. I wondered how the day-to-day work was really going to play out at Google.

"What about our raises?" I asked. "How much are they?"

"It depends on your performance," Nancy said.

"Can you ballpark it?"

"I don't know. It varies."

At my previous job, I only had to work in the office four days a week at most. Any day Gracie was home from school, I worked remotely so I could watch her while I filled out spreadsheets. It saved me a ton of money on childcare. For this new job, I'd taken a pay cut larger than the amount I would save on food, laundry, and gas. I needed that raise. But we didn't hear anything more about them during the presentation. At this point, I still felt that if I was patient and worked hard, things would work out.

In theory, when I started working at the Google offices, I thought I would also start going to the gym, but I could never bring myself to stay late or come in early, and during the day, I usually spent my downtime writing in a notebook. Now I would be eating healthier at least. I had lingering health problems that I'd been slow to address due to lack of insurance and not wanting to miss work. My new job had PTO, but I had to accrue it; we started with nothing. We also didn't have any dedicated sick days. If anyone got sick, they used PTO or supplemental leave days, which were somehow different and harder to use. The ticketing system looked like a homemade webpage from the late nineties. The process involved screenshotting an email and uploading the file to another leave management system. There were at least ten steps. We were told to make sure to do it well in advance in case the team in India was asleep when we filed the tickets. They also sometimes missed them, and we were told we would need to follow up the next day. If there was a problem with a ticket, you had to file another one to get someone to fix the issue.

After Nancy's presentation, we had security training and signed some paperwork. Part of what we signed was a nondisclosure agreement that included a forced arbitration clause in the fine print.

I dropped my new hire packet at my desk and organized my things. Tom, the lead of Clustering, stood at a desk by the back wall in the pod next to ours, coughing like he was about to die.

Everyone sat or stood at their desks in the glow of their computer screens; most wore headphones while they worked through their daily tasks. Almost everyone looked much younger than me. Small potted plants sat on the windowsills and the gray shelves separating the clusters of desks. Nerf guns lay scattered about. Big metal shelves covered with action figures and books on coding lined the main walkways. It felt like a copy of a copy of someone's idea of what a cool office should look like. The shades were drawn on most of the windows, presumably to prevent corporate espionage and to soften the morning glare.

After orientation, we went to lunch. Sam, Bobby, and I all agreed the job seemed shadier than we had initially thought, but we conceded that the free food was great. I had a sandwich and a salad. The three of us sat together and talked about our past jobs, why we'd taken this one, and how over-the-top the office was. We talked more about the interview process and how off-putting the recruiter had seemed to us in retrospect.

After lunch, it was time for the tour.

June 2018
OFFICE TOUR

Eva had just been promoted to trainer. Part of her duties included showing us around the office. Like almost everyone else there, she was at least ten years younger than I was. She led us through the three floors of the building on the north side of Penn Avenue, answering questions and offering advice. She spoke with a slight upper-Midwest accent that reminded me of *Fargo*. While we walked, she pointed out some areas that were under construction.

"I have no idea what's going on there," Eva said. "They don't tell us anything."

Some doors wouldn't open for our red badges, so we had to take a circuitous route. None of us was in a hurry. The two buildings that held the Google offices were both L-shaped, and there were elevators at the angles and on either end. Not quite a maze, but it felt disorganized. Double doors opened up into breezeways and more doors with blinking badge readers. A whole floor of the building on the other side of Penn Avenue was under construction; we could see the materials through the plastic hanging from the windows. Many of the windows on the other floors were still covered. Where we were, there were notes to clean the whiteboards before leaving the conference rooms.

On every floor, there were maps by the restrooms and microkitchens. I needed them because the conference rooms weren't numbered. Instead, they were all named after places in Pittsburgh, local celebrities, or, in the case of the floor where my desk was, rides at Kennywood.

We passed a cluster of walking desks that overlooked the parking lot of the strip mall surrounding the office. I could see the bar where my baseball league held its preseason meetings. I wondered how long it would take for this place to become run down like the strip malls on Route 1 in New Jersey that I remembered from my youth. What would have to happen for the office to turn into a Halloween store? For the new Anthropologie to become a massage parlor? For Jimmy John's and Starbucks and Panera to become a Cash for Gold, a check cashing place, and a Dollar General?

I thought about all this and the mystery surrounding our raises while we continued exploring the office. We went downstairs, and I got a coffee from the in-house coffee shop. I wondered how much the baristas made and if it was a good gig or not. Everything was free, but there was no tip jar. On a shelf next to the register, there was a record player and a small case full of pastries from one of the local bakeries—the gluten-free one that had popped up on Penn in place of an old art gallery. I didn't feel comfortable taking one; I felt like I shouldn't go crazy with the food too soon. The coffee, which was from Zeke's in East Liberty, another place that was relatively new, was delicious though. There was an exit by the coffee bar that led into a hallway and the offices for the University of Pittsburgh Medical Center (UPMC). They made prosthetics there. Sometimes I'd leave that way and walk down the hallway toward the Hamilton Street exit and see war veterans testing out new legs, or doctors carrying robotic arms.

Eva led us across the walkway in the sky—a glass and steel bridge over Penn Ave that connected the two Google buildings. We passed one of my daughter's friends' dads. I nodded hello, but he didn't respond.

Our daughters had been playing softball together for three years at that point. Record-wise, the team was pretty bad. I helped coach; I knew the fundamentals of the game, and I enjoyed teaching the players. The most thrilling thing was when Gracie did well and the team won. It also gave us a chance to get out of the house for games and practices and spend time together.

In the corner of the other building was a library and an outdoor porch with couches, where a handful of people sat and worked. Eva said she liked to sit out there when she needed a break.

On the tour, Eva said hi to a lot of people. She had started in the Pittsburgh office before HCL absorbed all the contractors and temps who were working on Google Shopping. There was enough work that we needed to be hired full-time, but we weren't official Google employees. I'd been in a similar situation before at Dick's Sporting Goods, where I'd worked as a copy editor with Quad Graphics. I had started there right before the company gave out a cost-of-living increase. My boss told me not to tell everyone else that I got the raise too.

"They've been here for a while," she said. "They might think it's unfair."

My pay went from $15.50 an hour to $16.10, an increase of just under 4 percent.

During that time, Jane was a mess. Gracie was with me five days a week, and for my job, I had to be an hour away in Moon Township for the entire workday. Four copy editors were crammed into a small office. The other workers went to HR to complain that I smelled like cigarettes because I smoked on the way to work, so I chewed nicotine gum all day and it hurt my stomach. We worked with physical, paper proofs, and there was no way to bring them home. So I had to arrange for a babysitter every day Gracie was with me. The one good thing about that job was that I could work unlimited overtime, but I had to spend that money on a sitter anyway, so instead of time and a half, I'd take home under ten dollars an hour. My old Honda was on its last legs, and I was pouring money into it to keep it running. I was worried it was going to die in the Fort Pitt tunnel or on one of the bridges. I was also getting killed with legal fees and lost wages for the time I had to take off work for family court. I had already run up thousands of dollars in credit card debt. After five months at Dick's, Thermo Fisher offered me my old job back. They paid better, and I could work from home whenever I needed to. Even if I was only there for the initial six-month contract, it at least would buy me some time to find something else. Six months turned into two years, and "something else" turned into HCL.

I don't know if I ever thought I would have a normal career path, or if I even wanted one. I wanted to provide stability for Gracie and to have enough time to pursue my writing in a meaningful way. But I didn't want to force the starving artist lifestyle on my daughter.

Eva led us down a hallway that was named after a local park. Each of the small offices was made up to look like a hollowed-out tree.

"We use these when we have to make private phone calls," Eva explained. "Sometimes if I'm not feeling well, I'll lay down under the desks and sleep." She pointed to a small, dark room with a recliner and a private docking station.

"We are allowed to use the nap rooms upstairs, but not for more than an hour."

"Nap rooms sound gross," I said. The thought of laying down in someone else's sweaty sheets who'd been at the office for days was unsettling.

The office had an in-house masseuse, massage chairs, a personal trainer, yoga teachers, and helpful receptionists at the desk by the gym near the back entrance. People kept their bikes locked downstairs if they rode to work. There was a doctor the full-time employees could see if they were sick, a hair stylist with reasonable prices, and even in-house dry cleaning.

"They don't want anyone to leave, do they?" I asked.

Eva laughed. "Nope," she said, and we continued up a small flight of stairs and badged in through another door.

I wasn't sure if Eva was considered a manager or not. She seemed very earnest, and I couldn't tell if she was really into Google or just good at her job. We passed the nap rooms and the library again and came to an open workshop in the back with 3D printers and other robotics stuff people could use to do their own work after hours. Then there was a big climbing net and some weird tubes like the kind they have at Chuck E. Cheese. This led into the music room and then the ping-pong room. I was worried I'd get lost one day and accidentally walk into a sex robot room. I had no idea where I was in relation to my desk.

We finally came out on an empty corridor lined with cubicles on both sides. When I saw them, I suddenly longed for my boring Thermo Fisher office out by the airport, where I'd watch the planes take off from the window seat I had maneuvered into over the years—my own space. My first day at Google wasn't half over yet, and I was already regretting my decision. I felt like I'd given my bank account info to a former Nigerian prince. All the window dressing was insulting for a job that paid $19 an hour. The perks were there to keep the high-priced talent happy and on-site. HCL used them to subsidize low pay.

"What's the one thing we should avoid doing to not get fired?" I asked.

"Don't take your laptop out of the building," Eva said.

Google couldn't or wouldn't let us work from home, which didn't make sense to me. But I figured I would adjust. After all, I lived close by.

We got back to our pod and took our seats. Tom from Clustering was still coughing violently in the corner. For the next few hours, I followed the directions Eva gave us, using Google's proprietary tools to curate catalogs. I reviewed an endless stream of products, determining a catalog's identity and removing incorrect products by clicking a square. I made sure the description, title, and image were correct. If they were wrong, I edited them. This improved the shopping experience for consumers. The work was dull. I had no idea how the algorithm produced the stream of products, only that it did and that my job was to edit them. But I assumed this would be only one part of my job and that I would learn more tasks soon.

While I worked, I saw a few full-time Googlers walking around in bare feet. One guy bounded past our desks, leaping and talking loudly; he may have been wearing a Bluetooth headset. Another person walked by with a tiny dog on a leash. One guy was carrying a Nintendo Switch and wearing those shoes with individual toes. I thought I recognized a couple people but couldn't place them.

Nearby, about two steps below our horseshoe, there was another cluster of desks. George, Alice, Carolyn, and Kate sat down there. They'd introduced themselves earlier, and George said hello again after the tour. George also spoke French and liked to post *Absolutely Fabulous* memes in the group chat. His dog, Kingston, sat under his desk. I had a difficult time keeping the three women straight. I was intimidated by how close the four of them seemed and how much smarter and more accomplished they all were. Carolyn wore a Devil Rays hat, big sunglasses, a tracksuit, and bright white Nikes—like a mob wife in witness protection. Alice's hair was cut short; she'd been through cancer treatment and had only recently returned to work. She made darkly funny jokes about surviving. Kate reminded me of a friend from high school who moved away when we were kids and who I hadn't heard from since the nineties. Alice and Carolyn both spoke German and English. Kate spoke Spanish and English. Bobby and I were the only two nonbilingual workers on our team.

I put on my headphones. Behind me, Tom kept coughing. He hacked intermittently until it was time for team building.

June 2018
TEAM BUILDING

We had team building for an hour each month. According to Eva, it was usually some kind of board game session or holiday-themed activity. They could have given us twelve more hours of PTO every year, but instead we got team building. This was one of our perks.

This month, all the HCL teams on Google Shopping had team building together; ten teams with about ten people each. Nancy had set this up so the new hires would get a chance to meet everyone. There were four new hires, the three of us on Catalogs and Cynthia, the new French speaker on Clustering. In the coming months, Cynthia and I would take smoke breaks together and talk about books. Her parents lived up north near Gracie's grandmother. She was very affable and knew more about literature than I did. The four of us who were new all got along well. We were a strong cohort. However, I had no desire to meet everyone.

Donald, the middle manager right below Nancy, was also there. He seemed awkward but harmless, oddly conservative for someone in his early thirties. Despite our lax dress code, he always wore a button-down shirt, dress shoes, and dress pants. He alternated between two expensive-looking briefcases that everyone assumed were empty. Someone told me he'd gone to an Ivy League school. He was a lead analyst, which meant he did administrative work and acted as another buffer between us and our team lead, the operations lead, and Google.

We went across the street to the baseball field at Mellon Park between Fifth and Penn. The infield was full of rocks and sloped down toward the left-field line. Summer rain had worn deep ridges between the pitcher's mound and the third-base line. Neither the city nor the local Little League association took good care of the field.

In shallow right field, my coworkers played cornhole and frisbee. George, who seemed like the unofficial mayor of the office, held court by the bleachers, and people gravitated toward him and his dog, Kingston. I scratched Kingston's ears and said hello on my way past but mostly kept to myself. The crowd made me uncomfortable. I was bad at small talk. I was not a fan of this type of forced camaraderie, but at least we were on the clock.

After a few minutes, Bobby introduced me to Darren, who he knew from high school. Darren commuted forty-five minutes to the office from Greensburg every day.

"That's a hike," I said.

"I get in early and beat the traffic," he said. He had a huge beard and wore boat shoes. Initially, I didn't trust him. But I figured I probably seemed like a weirdo to most people who met me, when I was just introverted and daydreaming or stressed out. Still, I didn't like the way Darren smiled when Bobby introduced us.

"How long have you worked here?" I asked.

"Since before HCL."

It seemed like a lot of people had been there for a long time, which I took as a good sign. I still hadn't met many workers closer to my age, though, which I took as a bad sign.

"What's the deal with our raises?" I asked.

"They're not bad," he said.

"What do you mean?"

"They're pretty nice," he added.

"The starting pay isn't great," I said.

"It'll get better."

Darren walked away toward another pack of workers. Or maybe I did. I don't remember—but the conversation ended with me feeling weirded out by how unforthcoming everyone was about money. For the rest of the hour, I sat in the shade and watched a group of people kick a soccer ball around in the outfield.

I thought back to 2003. I'd been in Pittsburgh for two years when I started playing summer league baseball. In the early days, my team occasionally practiced at Mellon Park. There had once been a middle school where the condos and orange houses now sat. Kids from the Boys and Girls Club used to run around in the field on the adjacent lot while we practiced. When we play games there now, the new office, the construction site of the future building, and all the condos butting up against the park make me feel claustrophobic. During the day, the shadows cast by the new buildings make it harder to hit, and the glare off the steel and the glass windows behind home plate make it hard to field. Night games aren't as bad, but the lights are too low, and it's tough to track fly balls. The surroundings don't make sense aesthetically, and the effect is disturbing—a baseball field treated like an afterthought, surrounded by a business park in a dystopian neighborhood.

During team building, I overheard Donald talking to one of my coworkers about the new building and the development behind it.

"It's amazing the progress the city has made," he said.

After an hour, Bobby and I walked back to the office with the group. We waited at the light. Traffic was backed up down toward the empty lot where the Penn Plaza apartments once stood. The city demolished them to create space for a new Whole Foods. Five hundred residents were forced out of their homes.

There was already a Whole Foods within walking distance.

"No one will talk about their raises," I said.

"Darren wouldn't tell me either," Bobby said.

"What's his deal?"

"Libertarian."

Back inside, I went over some training materials, played a few games of pinball, and killed time until it was 4:00 p.m. Until I was trained on more involved tasks, I could get a day's worth of work done in three or four hours.

Gracie had a softball game that night. I picked her up in Lawrenceville, made dinner, got her uniform ready, and then walked to the gas station for a Gatorade while she changed. Then we drove back down Penn to the ball field behind the one where I'd had team building. I parked between the new Google building and the new apartments. I helped line the field while Gracie played catch with the manager's daughter.

There was only one team in the league that was worse than us, and we weren't scheduled to play them again until the end of the season. The challenges of coaching a bad team were different than coaching a good one. With a bad team, it was easy enough to rotate positions and make sure playing time was even, but it was hard to keep everyone interested. We also ran the risk of losing our better players. But with a good team, I would have felt obligated to coach to win, which presented an inverse set of problems. Most of the parents were fine, though, which I was thankful for. During our games, the manager made all the lineup decisions, I coached third and waved everyone home. I tried to encourage our players to stay in front of the ball and reminded them how many outs there were. I was primarily concerned with how Gracie was hitting and that everyone had fun.

That night, we lost by fifteen runs, but Gracie got a hit. The full-time Google employee who I'd seen in the hallway earlier that day was there watching. He avoided eye contact with me and didn't say hello.

March 2019
NIGHTMARES, ROBOTS, AND PAIN PILLS

I woke from another nightmare, almost falling off the couch and gasping for air in the pitch black. I walked to the bathroom, taking tiny steps so I wouldn't tear open the stitches in my butt. I emptied the piss bag into the toilet, cleaned the blood off my ass, and put on a clean diaper. I hadn't shit in five days since the surgery, and my stomach was killing me. At least the catheter was coming out soon.

The living room smelled faintly of blood and urine. I heard screaming from outside and waddled over to the window. The neighbors across the street were fighting again. The door slammed and the woman walked up the street and waited until a black truck came and picked her up. I watched as the guy sat on his front porch, smoked a cigarette, and stared at his phone. I knew they had kids. Calling the cops on them would be a disaster, and most likely it would only make things worse. So from my window, I wished them luck and watched them like an alternate timeline of a life I might have had if I hadn't stayed sober and I had let Gracie's mom stay with me. Public fights and heroin and dealers coming and going. I couldn't imagine being in the middle of it.

I split a pill in half, put the big piece in my mouth, and let it dissolve on my tongue. I sank into the couch and put on *The Sarah Connor Chronicles*, the *Terminator* spinoff series from the early 2000s. The first time I had watched the show was in 2009 when I was working nights as a busboy. Gracie was almost two years old then, and her mom was coming unglued after several years of being clean. We were on the cusp of a long, expensive custody dispute. The economy was in the tank, and I couldn't find a day job with an English degree and no experience. I had a part-time job editing social media posts a few hours a week. All I wanted to do was write. I wanted to apply to grad school, but with Gracie, I couldn't leave Pittsburgh. Jane was on probation and couldn't leave the state, and I wasn't in a position to take my daughter and move away from her mom. Options were limited.

I felt like a loser. At that point, I'd been sober for almost four years and hadn't done anything with my life. The restaurant I worked at wouldn't even make me a waiter, and I was on food stamps. I would come home with a hundred bucks after bussing tables all night and sit and watch *The Sarah Connor Chronicles*. The show itself isn't amazing, but I found it at the exact right time in my life. I love the *Terminator* mythology, both despite the botched reboots and the abandoned trilogies and because of them. The depth of the letdown of all the sequels after *T2* was great. What should have been a science fiction powerhouse full of time travel, robots, artificial intelligence, and sharp social criticism turned into a string of mediocre action movies. But so much wasted potential devolving into a messy, nonsequential faux serialized movie franchise was still interesting to me. Maybe, on a subconscious level, I found the failure relatable. The unintentional gaps in the narrative allowed me to fill them in myself, and I watched *The Sarah Connor Chronicles* sympathetically. Compared to most of the movies they had released, the show was, at worst, solid B-movie fare and, at best, an excellent adaptation of the *Terminator* world. I connected with Lena Headey's portrayal of Sarah Connor as a single parent struggling to connect with her kid. In 2009, it was exactly what I needed to watch after work.

As I watched it now, though, high as fuck on the pain meds, I grew wistful and sad. I felt just as stuck in my job at Google as

I did when I was a busboy, writing bad short stories and editing Facebook posts during the day, worried that my daughter's mother would overdose while I was clearing plates off the table of some coked-out socialite or sex-pest quarterback. It was a good restaurant job—I made about as much hourly as I did as a contractor at Google. But I couldn't keep working nights if I ever hoped to gain full custody of my daughter. I'd be working just to pay for a babysitter.

I lit some incense to cover the smell in the living room and took a few unsatisfying puffs from a Juul I had bought to hold myself over while I was confined to my apartment. Then I went back to sleep, sliding in and out of vivid, opioid-laced dreams about cyborgs, the apocalypse, and work.

In these dreams I was trapped in the office—sometimes with Gracie, sometimes alone. The office had bunks in it, and I would wake up there, feeling lost. I would wander through the place. I'd see coworkers at their workstations or in their own lofts in what was a demented dormitory draped in bright primary colors. Everything I needed was there, but my badge wouldn't work when I tried to leave. Then I was being chased by hordes of the undead, or robots, or some other monster that I couldn't see. So I ran until I was out of breath and hid. When the coast was clear, I tried to escape, but every hallway and stairwell led to the same place. All the doors were locked, and the windows wouldn't open.

July 2018
LONG WEEKEND

On Fourth of July weekend, my girlfriend, May, and I were on our way to play pinball on the South Side when Gracie called from her grandmother's phone. When I answered, she was bawling. She struggled to catch her breath, gasping in deep sobs.

"My mom died," she said. "Come pick me up."

"I'm leaving right now."

I hung up, got in the car, and drove to Evans City, about forty-five minutes north of Pittsburgh. May's house was out of the way, so she came too. She'd only met Gracie a few times.

"She passed all her random tests," I said. "Hair tests, even. Can you beat hair tests?"

May said, "I'm so sorry."

"At least Gracie is OK."

I chain-smoked my way up I-79. May counted the change in the center console and organized the glovebox. I wanted to quit smoking, but that wasn't happening today. I thought back to all the times I'd driven up north when Jane and I were dating and she was staying with her mom. We were so young. I was less than a year sober and still in college. My whole life was ahead of me. Back then, she was so funny and happy. I didn't have a car at the time, so we drove around in her beat-up old Dodge hatchback and went to shows all over the East Coast. What should have been a fun fling went bad, though, when I wanted to break it off and she didn't. Then she got pregnant with Gracie.

It felt inevitable, but there was still no way to prepare for it. Ten years of struggle ended like it was nothing. Jane was gone. Some new, unknown pain would fill the empty space she left behind. The most antagonizing presence in my life, and the most important woman in my daughter's. Everyone thought she was getting her shit together. At least Gracie got to spend a few final days with her when she was clean and in good spirits.

Jane had overdosed the first weekend after completing her court-mandated drug tests. It was the first weekend I could no longer legally keep Gracie from staying with her overnight. Ever since I'd known her, Jane struggled to stay clean when there wasn't the threat of jail or rehab hanging over her. Looking back at all the times I had talked myself into letting Jane take Gracie even though I thought she might be getting high, I felt like I had dodged an even bigger tragedy. All the horrible hypotheticals I played in my head felt certain now. Gracie's early years were suddenly different. Her mom loved her. I never doubted that. But stories of weird houses and strange men and a mom who sometimes fell asleep for no reason filled her memories, and one day, she would put all of it together. Gracie picked up on things. She was sharp.

I was heartbroken for my daughter. I was also relieved, and I felt guilty for feeling that way. I was mad at myself and at Jane. I was mad at Narcotics Anonymous for stigmatizing harm reduction and Suboxone. I was mad that our country treated addiction like a crime and not a health crisis. People kicked dope and stayed clean. People with kids who needed help got on methadone and stayed alive. Jane couldn't. It wasn't fair.

My car bottomed out and scraped the gravel when I pulled into the driveway. Gracie was waiting outside with her grandma. I helped gather her things, and we headed back to Pittsburgh. Gracie sat up front with me, and I listened as she talked and cried. I tried to focus on her and not on the added pressure we would both be facing for the indefinite future, the way we'd have to completely restructure our lives. There were no immediate plans for a funeral, but Jane's family said they would set something up later in the summer.

"I didn't say goodbye," Gracie said. "Maybe if I had stayed up to watch a movie, she wouldn't have died."

I rehashed a lot of the talks Gracie and I had after her mom went to rehab and jail and then rehab again.

"Your mom was sick," I said. "It's not your fault. It's no one's fault. We all tried to help her. I love you so much."

"I should have gotten up earlier. I could have called an ambulance." Gracie's face was red and streaked with tears. "I saw her on the floor in her uniform. Her hair was pulled back for work."

Fuck, I thought.

"Honey, it's not your fault," I said.

"Why did she do it? She was better. I could tell when she was high, and she was better."

This had been going on since Gracie was two years old. I'd lived in constant worry about Gracie's well-being, with only minor reprieves when Jane was institutionalized or on probation and holding down a steady job. She'd keep up appearances just long enough that I would let my guard down. Then she'd relapse again, and it would be hard to notice right away. There was always a lag, and then it snowballed, and I'd find myself trying to figure out how I'd let it go for so long. It was such a sad story to piece together backward, using only chunks from court records and police reports and the stories Gracie told me. The court system had failed all of us. I was only able to get full custody in 2015, after Jane overdosed at the playground down the street while Gracie was her care.

Now it was over. The fentanyl-laced dope that had been killing people in the region had found its way into Jane's arm, or wherever she still had a good vein. Gracie wouldn't have to take care of her mom anymore or worry about needles laying around or worry about lying to me about what her mom had done or where they had gone. But now she didn't have a mom. Now it was a whole new thing to have to deal with. I was pretty sure Jane overdosed and died while her boyfriend was working the overnight shift at the furniture warehouse and Gracie was asleep in the next room. Or maybe Jane got high before heading out to work a double at the restaurant while her boyfriend was still asleep after working until 4:00 a.m. Gracie said she thought her mom was still alive when she saw her. She told me she had said "I love you" before she went to bed.

We'd have a lot of help in the coming months, but it was mostly just me and Gracie. That made life both simpler and more complicated. Our lives would be hard in ways that hadn't surfaced yet and that we couldn't prepare for. I'd figure out the logistics of everything eventually, but it was going to take time. Driving back to Pittsburgh with Gracie weeping in the passenger seat, even thinking about it was hard.

The next week was the Fourth of July. I had Thursday and Friday off from work, and I was entitled to three days of bereavement leave. When I went back to the office, Gracie went to stay with my parents in New Jersey. It was too hard for her to be in Pittsburgh. I tried to shuffle my schedule, and hers, to accommodate our new life as a two-person family unit. May was nonjudgmental and understanding, and I was grateful for her support.

Gracie and I didn't have much room at my place. Four months earlier, I had bought our house from my old landlord, but another tenant still lived downstairs. He was an independently wealthy consultant of some kind, and I could have charged him twice what he was paying. As it was, he paid $550 a month to rent the first floor of my house and have use of the backyard, the basement, and one parking spot. He was mostly quiet, a little twitchy. Because he had money, I naturally didn't trust him. I didn't want to be a landlord, but I didn't want to kick the guy out, money or no. So I just let him pay what amounted to less than half my mortgage and told him to start looking for a new apartment.

When I was back at work, I called HR and tried to change my health insurance, but it was hard to get a straight answer on anything. They sent me to the insurance company, who sent me back to HR. I filed multiple tickets. I thought about keeping Gracie on Jane's insurance until someone figured out Jane had died, but if I was jailed for fraud, Gracie could wind up in foster care. I tried signing up for CHIP, but my annual salary was $1,000 too high, I didn't have the proper documentation for my writing income and a number of expenses I'd written off for a book tour that was now cut short, and I had a home office that was being used less and less due to HCL's on-site only policy. If I had quit my second job, I could have qualified for an $80 per month reduction in my health insurance, but the job brought in $200 a month. I was right on the cusp, where I made too much to get any help and not enough to not have to worry about it.

If I could work from home, I could take a second job. If I couldn't, I'd have to try to work overtime. But without some flexibility, the situation would become untenable. I could live with no time or no money, but I needed one or the other. Right now, I was fucked. I was without child support, and I had to pay for Gracie's health insurance. I got the United Healthcare Gold Plan at work so her heart appointments would be more affordable. The out-of-pocket max was thousands of dollars, so I had to make sure to plan any big expenses before enrolling so I wouldn't have to worry about the deductible.

At our next monthly one-on-one meeting, I asked Donald, who had recently been promoted to team lead, if I could work from home.

"We don't do that," he said.

"Why? I need to be around my daughter. I don't need someone watching me do this work. Who do I ask? Who makes these decisions?"

"It's not something that's done," Donald mumbled. He didn't look directly at me. "I'm sorry."

"It's not your fault," I said.

We were sitting in the Raging Rapids room for our meeting. Donald was visibly uncomfortable, and I felt bad about how mad I was getting. He didn't have a say in our policies, but he also didn't seem to think there was anything wrong with the way things worked. At the end of the meeting, he told me that Lisa, the new lead analyst, would take over my one-on-one meetings.

I grabbed my laptop and went out to a balcony that overlooked the Target down the street. I felt defeated and alone. With Jane gone, I would now have to go to every event, every parent-teacher conference, and every doctor's appointment with Gracie. The Boys and Girls Club was relatively affordable, but it started to add up during the summer when Gracie stayed there all day. More importantly, I couldn't figure out how I'd support her emotionally through such a traumatic event. I blamed myself for all of it. For not doing my due diligence before I took this new job and for leaving a job that was flexible. Then I told myself I was lucky to have any job and that I was in a better situation than a lot of people. I hated feeling like a sucker. I hated that conditions at work were making my kid's life harder for no reason. But I needed the health insurance, and I couldn't quit without something else lined up. So I updated my résumé and started applying to jobs.

The next day, I went over Donald's head. I made an appointment with Nancy to work out a plan so I could better support my daughter, either by working overtime to ease the financial burden or working remotely so I could be physically present more of the time. Nancy told me Google wouldn't allow overtime, so I asked about working from home. TVCs from different vendors worked remotely. The internal Google websites said it was possible. Someone just wouldn't let us. Either Google or HCL had decided we should be on-site all the time. A manager had to put in a remote permissions request for us, and then Google could approve it.

Nancy said she'd ask Jeffrey, the deputy general manager who was never in Pittsburgh. In a week, she got back to me and said my request had been denied. I have no way of knowing if she ever asked him or if he ever asked whomever it was that made the final decision.

I'd been a contractor before with a few different companies and had always been able to talk to the person who made the decisions about my job. HCL said Google wouldn't give me more flexibility. Google's policy was that HCL had control over my working conditions and that I should go to my manager with issues. I was stuck in a loop.

———————

In late July 2018, *Bloomberg News* published "Inside Google's Shadow Workforce," which detailed the company's increased reliance on TVCs and the disparity in working conditions between the different tiers of its workforce. It laid out everything that I'd learned firsthand since I started working at the Pittsburgh office, and it included horror stories from other TVCs about sexual harassment, health care, and the feeling of being voiceless. As the authors put it, the contract workers were "a sea of skilled laborers that fuel the $795 billion company but reap few of the benefits and opportunities available to direct employees."

After the article came out, there was lively discussion in the work chat rooms and on our coffee breaks about working conditions. A few outspoken workers said it was about time someone wrote about the fake Google jobs we all had. The old guard at HCL in Pittsburgh knew what it meant to work *at* Google but not *for* Google. More articles about the inequity of the tiered workforce followed in the coming months amidst growing unrest at the company.

A lot of us were angry. We were angry at HCL and Google but also at the full-time employees who we thought were helping to perpetuate the divide between workers. The whole thing felt like a giant pyramid scheme. I decided to give it another few months to see if it got any better. Maybe I'd stay until I got my raise. I would try to power through until my life settled down, but I was still worried that I'd made a terrible mistake.

September 2018
QUESTION OF THE DAY!

Occasionally, one of the trainers would write an innocuous, usually funny, question on the big whiteboard on the wall between the pods on the north side of the office. They reminded me of *Buzzfeed* quizzes. They'd ask about vacation destinations and *Game of Thrones* characters or things like, "What was your favorite childhood Halloween costume?" One day someone wrote, "If you could have one wish, what would you ask for?"

The exercise was mostly vapid, like team building. I felt bad for being so cynical that I couldn't enjoy minor pleasantries like this at work, but I didn't participate unless it presented a real layup for a joke. Usually the answers were ironic, but on this day, someone must have been in a mood. The first response, written in George's handwriting, said something about socialized health care. Then someone wrote about not getting time off when the office was closed. "Permanent employment with Google" was on there. As the morning wore on, a laundry list of left-leaning workplace grievances covered the surface that normally held nothing more controversial than our favorite nineties movie characters. I didn't write anything. The board was erased after lunch.

A few weeks later, a *New York Times* report about former Google executives who had been accused of sexual misconduct made the rounds on our work chat. The company had paid millions of dollars in exit packages to executives accused of harassment, and it had stayed silent about their transgressions.

Andy Rubin, the creator of Android, received a $90-million payout when he left the company in 2014. There were others as well.

The company that wouldn't hire us as full-time employees could afford to give Rubin $90 million. I shouldn't have been surprised. Google's PR team had done well, and at that point, the company still had a pretty good reputation. It was starting to crack, though, and I felt dumb for believing Google was a different kind of company.

Just days after the settlement was reported, we saw flyers for a walkout. "TVCs Welcome!" was written in big letters across the top.

The Google walkout was organized primarily by full-time employees in San Francisco in response to Rubin's exorbitant severance package and the company's failure to hold him and others accountable for sexual harassment allegations. There were about seven organizers, most of them women. The workers made five demands: an end to forced arbitration; a commitment to end pay inequality; a transparent sexual harassment report; an inclusive process for reporting sexual misconduct; and an elevation of the chief of diversity so that they answered directly to the CEO, combined with the creation of an employee representative on the board. The walkout received a great deal of publicity. Over twenty thousand workers participated worldwide. Google agreed to end forced arbitration for full-time employees in cases of discrimination and sexual harassment. The company did not end forced arbitration for TVCs, and Google did not address any of the other demands.

The walkout wasn't the first time workers had organized at the company. In April 2018, over three thousand workers signed a petition against the Project Maven drone program. The petition was in response to engineers' labor being used for military surveillance and drone warfare. The gig workers tasked with image labeling, which helped the AI discern satellite photos, faced a more traditional labor problem. According to the *Intercept*, "Outsourced crowd workers were tasked with providing the initial image data labeling—correctly identifying parts of an image—that allowed Google's artificial intelligence program to tell buildings, images, trees, and other objects apart." The article reported that these workers were paid as little as $1 an hour. Google eventually stopped work on Project Maven.

In August 2018, workers also protested Google's involvement in Project Dragonfly, a censored search engine that complied with China's state censorship laws. According to the *New York Times*, 1,400 employees signed a letter objecting to the project. Almost a year later, Google reportedly stopped working on the search engine.

The November 2018 walkout was the result of years of institutionalized sexism and discrimination that had culminated in the Rubin severance. It coincided with an increased sense of responsibility among a large section of the workforce that, among other things, their labor should not be used to kill people or perpetuate racism and state-sponsored censorship. Trump's election had also increased the urgency some workers felt about taking action.

At our offices, none of us really knew anything was being planned until we saw the flyers laying around the office days before the event. That TVCs were not usually looped in about any of this stuff was not surprising, given the way Google siloed contractors and full-time employees and ramped up its contracted workforce during this time. We couldn't act in solidarity with full-time workers if we weren't allowed to talk to them. Our managers also portrayed the full-time employees as privileged workers who thought they were better than us. For example, they'd tell us that Googlers would complain about us sitting in the same spot in the office for too long, or that we made too much noise walking past their workstation, or that when they stayed for dinner, there wasn't enough food because we had eaten at the office too. Mostly, they were just workers who were usually unaware of how little we made or the fact that we didn't get sick days or the same good health insurance they had. So a lot of us, myself included, had the idea that the full-time employees thought they were too good to slum it with us and build relationships, and, I would imagine, some of them assumed that was how we felt and so they didn't reach out, which in turn perpetuated the split. There were times when an elitist, out-of-touch paternalism showed in some of the full-time employees at the office, but the vast majority were kind and helpful. Some of the most radical and empathetic people I've ever met work at Google.

And in the years after the walkout, they continued to fight for equality and the just use of their labor.

October 2018
ALL-HANDS MEETINGS, SICK DAYS, AND A NEW LEAD ANALYST

Nancy joked about the lack of upward mobility and high turnover when she announced yet another new lead analyst of Catalogs in one of our "all-hands" meetings. Lisa was leaving, and Laura was taking over for her. She was my third lead analyst in three months. She was equally nice, and she also complained about the lack of transparency and opportunities to advance within the company. As it was with most of the lead analysts who had been promoted from within, Laura genuinely wanted to help me and my teammates, but in her particular position, it was difficult to make our work lives any better.

"We all know there isn't much room for advancement here," Nancy said, "so Lisa is leaving us to pursue other opportunities."

This was, of course, the opposite of what Nancy had told us in our interviews and what Stephen had told us during the recruiting process. It was bizarre. No one was accountable to anyone but Google, and none of the workers were allowed to talk to Google supervisors. I tried to find the humor in all of it. I stopped attending all-hands meetings, and I stopped going to team building. No one said anything.

I came into work one morning, and Tom was coughing up a lung. It was so loud I could hear it through my headphones.

In our stand-up meeting that day, we were told to sequester ourselves if we were sick.

"Try to find somewhere out of the way to sit and avoid close contact with FTEs," Donald told us. "Make sure you cough into your elbow."

This was the first time I spoke up in a meeting.

"How is this our policy?" I asked.

Donald stammered. He once again said it was out of his hands.

"I understand it is not ideal," he said, "but this is just how corporations work."

This man was in charge of real people with real problems. I wouldn't make fun of him for carrying an empty briefcase to work or for dressing like a participant in the Brooks Brothers riot, but I didn't like this answer.

"How can you say that with a straight face?" I asked him.

"We have PTO and supplemental leave days."

"We have to use them when the office is closed."

"There's nothing we can do."

"Who can do something?"

No one answered this question.

After an awkward transition, the meeting went on, and everyone sat quietly through the rest of the weekly updates. Our counterparts in Kraków watched in silence over the video monitor on the wall.

On the way back to our desks, Mark pulled me aside. He had a cop haircut, and he dressed like an aging skater. He looked exhausted. We stood near some ornate potted plants on the walkway above Penn Avenue.

"I used to get so angry," he said, "but I realized it was never going to change. I have to change. HCL won't. It's about how I deal with it."

"Thanks," I said.

Mark wandered off to work somewhere else. I stared out the glass windows at the cars passing below me, merging into one lane in advance of the torn-up sidewalk, and I thought about what he had said. If I'd had any savings, if I hadn't been solely responsible for Gracie's well-being, or if Jane hadn't died, I would have left on the spot. But now Gracie was depending on me for everything. I had made it through tough times before as a parent, but it seemed so daunting now. The health care system seemed impossible. I kept worrying I was forgetting about one of her doctor's appointments, and that if I missed one, I'd have to reschedule for months down the road. I felt like I should have had my shit together by thirty-eight. I was neither able to provide adequately for my child nor spend time with her. I could sense Gracie becoming more withdrawn, but I was so tired by the end of the day, it was all I could do to feed us both. I felt isolated and stuck.

In my mind, the job with HCL might have been tolerable in the short term, but I was falling apart. If they would have let us work from home or find some of the same work-life balance offered to Google's full-time employees, it might have been a decent job until I found something better. It might have been easier if I hadn't left a job that already paid better and gave me more time off. *I should have stayed until my contract ran out,* I thought again and again. I was always distracted at work. I had a mountain of administrative shit that I had to do for Gracie's school and medical care. She was struggling, and I didn't know how to help. I was barely making enough money, and the raises weren't happening. I felt trapped.

Each day was worse than the one before it. I couldn't stop beating myself up. Then I got mad at myself for beating myself up, which didn't help. I was mad for rehashing the same problem over and over for months. And despite the work I was doing to actively get out of the situation, I felt like a huge failure who was letting his child down when she needed him the most.

After lunch, I went to get a coffee and ran into George. He had his dog with him, so I joined them for a walk by the development out back.

"Tom sounds like he has the plague," I said. "He should stay home. I can't get sick."

George laughed and calmed me down.

"This place is not what I expected," he said, "but it allows me to conduct independent research and mostly do my own thing. I try to look at the positives."

"They take it so seriously but still won't pay us."

"I hear you, brother," he said. "A lot of us went through that phase where we wanted to change things, but we've all settled into what we have now."

George was a reassuring presence in the office. I wasn't friends with anyone but Bobby, who I ate breakfast with, and Cynthia, who I took smoke breaks with. On most days, I came in earlier than Sam, so our breaks didn't line up. The people I started with were all new and weren't in a position to offer advice on how to get by in this job. Alice, Kate, and Carolyn still intimidated me. I felt out of place. That feeling was exaggerated by how much older I was than my coworkers and the way management behaved. There was a lot of gossip.

"My daughter's mom overdosed on heroin. She died," I told George. "I don't know how I'm going to afford being a single parent, and I don't get any time with her because I have to work so much."

I'd overshared awkwardly with someone I didn't know that well in a professional setting. I wanted to hide.

George took it in stride and seemed to genuinely care.

"There are twelve-step groups out there," he told me. "I've heard they can be helpful."

I thought this was George subtly saying he was in recovery, but it wasn't. George was not in recovery, nor was he an alcoholic. He just figured if I was struggling with the death of an addict, Al-Anon could help. In part, because I thought he was discretely telling me he went to meetings, I immediately trusted him. I don't even trust most of the people I know from AA, but George seemed like he had his life together in a way that I didn't. Everyone asked him for help when there was an issue with work. He kept things light, and he went out of his way to make the workweek better for his teammates, especially me, Bobby, and Sam.

I knelt and petted Kingston. For a second, I felt guilty for complaining while I was outside near a goldfish pond, petting a dog on the clock.

Back inside, George went to his desk, and Kingston laid down next to him. I played pinball and then tried to write for the rest of the day.

A HISTORY OF HCL ON-SITE AT GOOGLE IN PITTSBURGH AND LESS THAN I PERCENT RAISES

Later that week, I ran into Kate in the laundry room at the office. She had originally come to Google through a different temp agency, and she had been there longer than anyone. She gave me more details about HCL moving in and taking over the temps.

Before HCL had employed anyone in Pittsburgh, all the Google Shopping workers had been employed by different temp agencies that paid them anywhere from $18 to $30 an hour, usually without benefits. To be considered an independent contractor, a worker can't hold the same job for over two years; if they do, they're considered an employee of the client company. Companies get around this by ending contracts before employees reach two years or by making workers take a furlough for a month. In 2014, HCL took over all the temps under one contract to put another layer between the contractors and Google and to allow everyone working on Google Shopping to become a permanent, full-time employee with health insurance. This meant Google wasn't subject to the joint employer laws and the length limits governing worker contracts. HCL made a litany of promises to the workforce when they took over.

Around this time, there was some talk of unionizing with the cafeteria workers, but it fizzled. For some workers, HCL wasn't ideal, but it was a slightly less tenuous existence than they'd had before. Over the course of hiring people, there was a dramatic difference in pay. The first workers HCL hired were paid under $15 an hour, half of what some of the longest-tenured workers made for the same work. Google employees doing the same work made upward of ten times that. HCL took away free parking and a few other minimal perks some vendors had negotiated for with Google on behalf of their workers. Some people quit when they could no longer work four ten-hour shifts and get Fridays off. When they left, they were replaced with cheaper, usually younger workers who were often attracted to the job because they thought they were getting a foot in the door at Google.

Kate also told me about how she had developed new tools and processes only to see Jeffrey put another HCL worker, one of his friends, in charge of the team she had created.

"That's when I stopped caring," she said. We watched the time tick down on the dryers while clothes tumbled inside them. "I used to complain a lot, but I lay low now."

"This is the first time I've been lied to by a recruiter."

"Lots of us have."

"I don't trust Nancy."

"You shouldn't," Kate said.

"They told me I'd get back to making over $50,000 in a year or two. I'm already applying to other jobs."

"Usually our raises are under 1 percent."

"How is that possible?"

Kate looked apologetic.

"Hang in there," she said. "Make the most of it and keep looking. I'll see you back at the pod."

"Thanks," I said. I took solace in finding someone who would at least be up front about money and wouldn't bullshit me.

Kate left. I sat on the counter in the laundry room for a minute while full-time employees passed by on their way to the gym. A flyer on the wall gave tips on how to talk to TVCs, urging Google employees to make an effort to be more social with those of us who wore red badges. After the flyer had gone up, a few random people sat down with me at lunch and tried to start up conversations. It was uncomfortable. I was ashamed of my job. I was also ashamed at how excited I'd been about it before I had started. I hated myself for being so stupid. I hoped Gracie didn't realize what I was going through. I tried to stay positive for her and remain supportive, but it was hard not to betray how sad I was most of the time. I didn't see how I would ever be able to send her to college or give her the kind of security I had as a kid, and I feared I was giving her something else to worry about on top of losing a parent, which I couldn't fathom.

I took my load of laundry out of the dryer and began to fold it. The repetitive movement was calming. I took my time. I still had two hours left in my day, but there was nothing pressing going on across the street, upstairs, or in our pod. Our whole team was built on the impression that we were working for Google. There may have been a time when TVCs had been converted with some regularity into full-time Google employees, but those days were gone.

I couldn't cut myself any slack, regardless of how bad the economy was. The idea of working hard and proving myself was bullshit. I felt like there was nothing I could do to improve my time at this place. The best I could hope for was that a prospective employer would believe I had actually worked for Google and give me a job on fake credentials.

I started applying to ten jobs a week. That cut into my writing time, but I was never going to be able to write if I was too stressed about money and family, and I saw no end to that stress if I didn't get a new job. There was no way I could tolerate this place unless something changed. Alphabet is worth $1 trillion. HCL is worth $10 billion. Surely, they could pay us enough money and allow us enough flexibility to live comfortably in the communities where politicians and developers had courted tech workers and driven up the cost of living.

Around the office, I heard conversations from full-time employees who had moved to Pittsburgh from Seattle or San Francisco. They were amazed at how inexpensive it was.

"I bought two houses," someone in the lunch line said one day.

"It's great to have the passive income," said another. "I'll keep my house here even if I move back."

I felt resentful whenever I heard these kinds of discussions, and my coworkers and I would generally laugh about how out of touch the Google workers were. It wasn't really their fault, and of course they deserve to be paid what they're worth. None of them realized how little we made relative to them, or to the value of our own labor.

Whenever I overheard discussions about investment properties, I thought about the guy living in the first floor of my house, and how I just wanted him to move out so Gracie would have enough room for sleepovers. Now I probably couldn't afford for him to leave. It all made me feel like a leech and a hypocrite and a bad parent.

October 2018
TVC HACKS

I got even more depressed as the days passed and the leaves changed color. Each workweek was worse than the one before it. I also started to worry about the upcoming holidays. I had already taken a couple days off to do a reading in Chicago in mid-October and another day off for parent-teacher conferences. Gracie also got sick one day, so I had burned more PTO taking her to the doctor. I couldn't accrue enough time off to cover the time between Christmas and New Year's, and it wasn't even Halloween yet. I gamed the system as much as I could, but it got difficult when Gracie and I had more than one appointment during the same week. I was constantly pressed for time and couldn't get any work done around the house. There were brief flashes of time when I thought I could gut it out, but then Nancy would write something in the chat like, "Don't forget to check out the new ice cream sandwiches in the coolers!"

I could not pay my bills with ice cream sandwiches. I couldn't even eat them because my teeth were so bad from not having dental insurance for most of my twenties. I was coming unglued. I developed a perpetual headache.

What made it worse was that I was constantly killing time at work, not doing anything, so I could later flex the time off. I stayed around the team average for productivity so I wouldn't stand out. I tried to work on my own writing on the clock, but I couldn't focus. No matter what I chose to work on, all I could think about was what I wasn't doing. If I was doing my actual job, I hated it and didn't want to do more than the team average, so I'd play pinball; if I was playing pinball, I was being frivolous and had to apply for jobs; if I was applying to jobs, I should have been writing; if I was writing, it was garbage, and I was never going to make it. And even when I did spend time with Gracie, I was usually too exhausted to really engage. These feelings heightened as the days got shorter and the sun set before 5:00 p.m. I was always tired, but I couldn't sleep.

When they refilled the microkitchens, I would stock up on the most expensive items. I didn't even like a lot of the stuff, but I kept my desk drawer filled with fancy snacks I would never buy for myself. Most days, I brought Gracie home a bag of chips and a Sprite as a treat. The random free bags of dried fruit, bottles of caffeinated flavored water, and strips of artisanal beef jerky didn't really save us money unless I substituted them for other items I would normally buy.

Whenever possible, I ate three meals at work. I started staying for dinner a couple nights a week. Sometimes I'd go home and see Gracie after school, get her settled, and then go back to the office so I could eat dinner and flex more time so I could leave early later in the week. On the weekends, I subsisted on spaghetti and pocketed office snacks. I got rid of my cable, cut back on cigarettes, and switched to a cheaper phone plan.

Nancy was never in the office, Donald seemed too timid to do anything, and I didn't respect either of them. I started taking chances. My teammates taught me to exploit gaps in the rules and create time for myself. In the morning, I would wait with Gracie until the bus picked her up at 7:15 a.m. Then I would drive to work and park in the garage by 7:45, before most of the full-time employees came in. This allowed me to get to my desk before any of my bosses showed up. I ate breakfast with Bobby at 8:00. At 10:30, after our morning meeting, I would move my car before my three hours of free parking had expired.

When I did that, I would stash my laptop in a locker across the bridge in the newer building before walking to the garage. I'd move my car to the street parking behind the building, put my laundry in, run upstairs, and make sure Donald saw me walking around, then I'd move my car again, this time back to Hamilton. I'd check for any work messages right before I left and as soon as I returned. I never missed anything important, and no one ever said anything. I got all my work done well and on time. It was depressing how important these petty acts of defiance became to breaking up the workday's dull ache.

This was my fourth office job. I'd worked on a farm, in a warehouse, at a lumber yard, and in coffee shops, restaurants, and bars. I'd landscaped, painted houses, and delivered pizzas. I'd never been infantilized and condescended to the way I was at HCL.

————

Bobby and I went down to breakfast before it got too crowded. We brought our laptops so we could presumably work during breakfast.

"Maybe we should start a union," I said, setting my tray down and sliding to the end of the booth.

"Yeah, right," Bobby said.

"Maybe I'll search 'union' on my work laptop so they'll fire me illegally. At least I'd get unemployment."

We laughed. Bobby said he only planned to stay a year so he could boost his résumé. I had already spent years doing more complex e-commerce work. I wasn't even gaining relevant experience.

I researched local white-collar unions but found nothing online. All kinds of public sector office workers had unions.

As an arbitrator for the state of New Jersey, my dad had settled strikes for hundreds of white-collar unions over the course of his career. But the private sector was a different animal.

I asked my dad if he knew anyone in Pittsburgh who could help me. He said it would be a conflict of interest for him to give me names. He had to serve as a neutral third party when he settled disputes, and it could look bad if he actively helped me.

"Good luck, though. It sounds like you need a union," he added. "Don't get fired."

"Thanks."

Unionizing was just a daydream. I knew we could theoretically form a union because we were permanent, full-time employees of HCL, but I didn't know if an international union would even be interested. It didn't make sense to unionize without an international union's help and the infrastructure and training they could provide. I didn't even know what that would entail. I knew about Pinkertons and Carnegie and the Homestead Strike. I knew about signing cards from watching *Norma Rae*, and about strikes and negotiations from my dad's job. I'd studied labor history, but there wasn't much instruction on what was involved in forming a union other than going on strike, which wasn't really an option for us. I knew Ronald Reagan busted the air traffic controllers' union in the eighties and that the working class has suffered ever since his presidency. Unionized workers had more leverage than individuals did, and they used it to demand better working conditions typically established in a collective bargaining agreement. My dad's job, in part, was to step in when the two sides couldn't come to an agreement and help them negotiate a fair contract, but he usually worked with teachers' unions. In the private sector, management doesn't even have to go to arbitration over a first contract. They can refuse to negotiate, and they aren't even punished for it. They can claim negotiations are at an impasse, and workers have almost no recourse. Some contract negotiations take years, and only about half of the first contracts ever get ratified. My basic sense was that unions helped workers. And we needed help. But I didn't know how we got from the point we were at to the point where we were signing cards or negotiating a contract.

October 2018
HALLOWEEN

Before Jane died, Halloween had always been a holiday she and I did together with Gracie. The last couple years, when Jane wasn't in rehab, we went out near her grandparent's house in Shaler. Gracie bounced around a lot when she was with her mom, but they mostly stayed with Jane's grandparents, who lived in a split-level in a quiet neighborhood.

Every Halloween I can remember, Jane was high. Cigarettes burned down to her knuckles while she sat on the porch steps and passed out candy. She wavered down the street full of monsters and superheroes while Gracie and I went ahead and rang doorbells with her friend, JR, who lived in the neighborhood. JR was a rambunctious kid a few years older than Gracie. His mom, Susan, drank a lot of wine and was always tired.

I had two costumes, a cardboard cutout blue Pac-Man ghost and a Ronald Reagan mask, that I rotated every other year. In 2018, it was President Reagan's turn. I put on my second-hand suit that I wore to weddings and funerals and an American flag pin. Gracie went as a zombie ghost. That year, we didn't go to Shaler. Instead, we drove to Morningside, where Gracie's second cousins lived with their mother.

"I miss JR," Gracie said.

"I can still take you over there if you want to hang out."

"I don't think he likes me anymore."

"Why is that?"

"He's mad I left."

After her mom had been arrested for the last time, Gracie left Shaler and started living with me full time. Her great-grandmother moved into a home in Wexford and was almost ninety years old. Her great-grandfather had passed away several years earlier. They sold the house.

"Maybe we'll go visit sometime," I said, but I doubted we ever would.

In Morningside, Gracie's cousins' porch was cluttered with old bikes and scooters. There were Halloween decorations up in the windows that looked like they had been made by the kids at school, and two jack-o'-lanterns on the steps with candles flickering inside them. Gracie and I stayed in the doorway and waited for the kids to come out. Morningside went all out with their Halloween decorations. There were multiple animatronic vampires, haunted houses kids could walk through, and lifelike movie monsters and zombies set up in scenes across front yards and porches. Families came from all over the city.

We trick-or-treated, and it was a nice night. No one nodded out on a neighbor's lawn or started a fight because a mom smiled at me while she was handing out candy. I took some pictures and joked about inspecting the candy so I could get a good cache of Reese's Cups for myself.

After Gracie's cousins went in for the night, we stopped in to see my friend Bruce and his wife, Jamie, at their house down the street. Bruce and I had gone to grad school together. He was one of the most stand-up guys I knew and a great writer. For a while, we had exchanged short stories when it seemed incredibly important to publish them all the time. We were both trying to figure out how to balance a creative life with financial stability. He worked as an adjunct at Pitt and as a personal trainer. I was so drained by Friday that I rarely left the house on weekends. Maybe I'd go to the occasional reading or literary event, but since I'd started the job at Google, I felt like a recluse. It was good to stop by and say hello and have a smoke on the porch while Gracie snagged the extra candy from the living room and played card games with Jamie.

I heard Jamie say, "I'll teach you how to play poker," as she shut the door, leaving Bruce and me alone on the porch.

"How's work?" I asked him.

"I love the kids but hate having multiple jobs," he said. "How's Google?"

"Really disappointing."

I tossed my cigarette toward the gutter, then I remembered Bruce's neighbors had complained about cigarettes in the street, so I walked onto Chislett, grabbed the butt, and put it in my pocket. I walked back to the porch as Gracie was coming outside.

"It could be worse," I said, and Bruce agreed.

We made plans to get lunch but didn't see each other again for a while.

I took Gracie home. I was glad she was able to get out and get a bunch of candy like a normal kid and also hang out with a woman I admired. I didn't know how much longer she would be trick-or-treating. I thought it might be the last time she'd go out. I was glad we got to spend this Halloween together before she started going to parties with other teenagers and wanted nothing to do with me. It would be a normal, happy Halloween memory. I was almost teary over it. It felt weird that it felt so good to have a regular holiday like this. But it was only normal because Gracie didn't have a mom anymore, and I didn't know how I was supposed to process that.

November 2018
GOOGLE WALKOUT

I called off on the day of the walkout. HCL said we could participate if we made up the hours we missed. This defeated the purpose. It wouldn't be a walkout if we didn't withhold labor. In the aftermath of the walkout, Google appropriated the organizers' messaging, publicly supported them, and vowed to do better. They claimed they wanted to figure out a solution to the problems workers were raising.

In the following months, though, some of the women who had organized the walkout were fired, and some quit under duress. Google also hired an anti-union law firm and started a campaign called Project Vivian, which according to *Vice*, was designed to show workers that unions were not cool.

The walkout's effectiveness is debatable. Organizers and participants brought attention to the tiered workforce, the pervasiveness of sexual harassment, and the discrimination that prevailed in a company that was supposed to be one of the best places in the country to work. They got a lot of press and drew attention to important issues. Twenty thousand participants is a lot, and it took courage from every one of them. The walkout was a catalyst for a tech activism revival in the United States. But it felt like a missed opportunity to push things further. After the walkout, people were burnt out, and there was disagreement about what direction the newly organized workers should take next. At the time, the most vulnerable workers at Google still didn't have much of a voice in the organizing efforts.

If a significant number of Google full-time employees went on strike for an extended period, they could demand almost anything. They have an unbelievable amount of leverage with the company, and most of them make enough money that they could probably hold out for a while. They were not done organizing, though, and the fight certainly wasn't over.

The year after the walkout, Google TVCs wrote an open letter to Alphabet, demanding an end to the tiered workforce. They asked for pay and benefits on par with full-time employees, a path to permanent employment, and access to the same company-wide information available to people who worked full time. In the face of growing unrest among workers, Google required staffing agencies to pay TVCs a minimum wage of $15 an hour and offer health insurance, parental leave, and eight dedicated sick days. Agencies had to meet the pay requirements by January 1, 2020, and the rest of the requirements by January 1, 2022.

HCL is still trying to avoid giving workers at the Pittsburgh office their sick days.

December 2018
CHRISTMAS

In December, Nancy suggested HCL workers go on one of the internal message boards and offer our services as professional gift wrappers to the full-time employees. She said we could charge $15 an hour, which, coincidentally, was the starting wage of our lowest paid coworkers. There was a wrapping paper station near the bridge connecting the two buildings, which was a nice touch. Many of us had packages delivered to the office, especially those of us with children who wanted to keep gifts from Santa a secret.

Eva was asked to plan the holiday party, which had historically been held at a restaurant in town and funded by HCL as an end-of-year thank you to employees. I felt like a better company would have given bonuses, but a party was better than nothing. I wasn't sure if I would attend. I still didn't have many friends at work, and I was not one to usually socialize with coworkers. Later that month, the party was postponed indefinitely.

I kept my personal life separate from work as much as possible, but I did once take George up on an offer to visit him and his husband at his husband's job; he was a regional manager for Dave & Buster's, and they had opened a new location in the suburbs. The traffic outbound through the tunnels was always awful, but I said I would go. It was a nice activity for me and Gracie to do during the week. Gracie spent her free tokens and then another $20 in half an hour. Then she started yelling about not getting enough tokens and wanting to play more games. We left unceremoniously.

Now that softball was done for the season, I was struggling to find activities Gracie and I could do together. I was busy with work and job hunting and writing, and she spent her after-school hours at the Boys and Girls Club. I was always exhausted, and she spent most weekends with her grandmother in Butler.

We were approaching her first Christmas without her mom. I somehow had already spent too much money on gifts and yet was also way behind on my shopping. I didn't want to have anything sent to work; I thought it would jinx my ability to find a new job. I tried not to use Amazon, which took some effort but was not impossible. I bought Gracie sneakers and a Nintendo Switch, some gift cards, and a bunch of forgettable items she either left in the packaging or immediately lost. I was grateful she had something to open on Christmas Day, though.

It was the first year I wouldn't be playing Santa. I rarely exuded holiday cheer but doing that had been one of my favorite things as a parent to a young child, along with wrapping secret Christmas gifts and then waiting to put them under the tree until Gracie fell asleep. The look on her face when she woke up and wandered into the living room warmed my heart.

When Gracie was eight, she spent the week before Christmas in the hospital with endocarditis. I suspected it was caused by her mom leaving needles around, but I couldn't prove it. It was terrifying because of her heart condition and the potential for a deadly infection. It was the hardest week of my life.

The following year, I house-sat for the novelist Chuck Kinder and his wife, Diane Cecily, over the holidays. Gracie and I spent a month hanging out with their cats and exploring their Pittsburgh-famous house on Wightman Street. That year, I was in between contracting stints at Thermo Fisher and working under the table for a realtor. Money was touch and go. My unemployment had run out, and I felt like even more of a middleman than I had working for HCL. I was doing some freelance work, but it wasn't steady. Jane was in rehab.

I hadn't had Chuck as a teacher when I was an undergraduate at Pitt, but he was immediately warm and friendly to me when we were introduced by mutual writer friends. They vouched for me, so I was OK by him. Before he died, I went to his last public reading. It was at an art gallery up on Penn; a few years after the reading, someone bought the place and turned it into a smoothie store. It was the only literary event I've ever been to that was attended by all the different Pittsburgh writer cliques. People who usually had open disdain for one another put it aside to see Chuck read one last time. That night, I met my friend Taylor. She read first and stole the show. We were introduced, and she's been one of my most important writing friends ever since. She also introduced me to May, years later. Chuck had a way of bringing people together.

After that reading, I went to my first and only literary party at Chuck's house. It wasn't exactly *Wonder Boys*, but it was still great. He said I was always welcome there and that I could bring Gracie over to see his cats whenever I wanted.

Chuck also said that if I house-sat and took care of his cats, he would read a copy of my novel and give me some feedback. At the time, it was getting rejected everywhere, but Chuck thought the manuscript was great. He liked everything and was always so kind, but it still felt good to hear that from someone so historically embedded in the Pittsburgh writing scene, a scene that I had struggled to find my place in for a long time.

House-sitting at his place, having all that space and all that time to spend with Gracie over the holidays, was the highlight of a brutal year. My family came to visit, and we all watched the Chargers lose a game that would have sent the 8–8 Steelers to the playoffs. Or maybe the Chargers won, but the Steelers needed them to lose. I forget. I wrote so much during that month. Snooping around in the basement, I found piles upon piles of Chuck's old manuscripts—towers of paper on old metal shelves. Gracie loved it there too. She loved the cats, and she was a favorite of Chuck and Diane until they eventually moved to Florida.

That Christmas, I felt a longing for a different past, somewhere in a parallel universe, even though I probably wouldn't have wanted it anyway. When I had gone to graduate school at Chatham, I thought an MFA would allow me to make a living teaching writing. My vision of Chuck Kinder and the mythology of the tenured writing professor was misplaced, though, especially in 2008, when I applied to MFA programs as the father of a one-year-old. The school told me they couldn't give me funding right away because at 2.95, my undergrad GPA was below the 3.0 threshold required for the limited fellowships, but if I did well, I'd have a chance to apply for aid down the road. The next year, they said I still wasn't eligible, but I decided to stay and finish my thesis, which turned into my first novel. I left with over $30,000 in debt and a terminal degree that qualified me to try to scrape together a living as an adjunct at five different schools. The most important professor and mentor I had in graduate school commuted from Connecticut to teach twice a week. Looking back, I should have seen that as a warning sign about my own prospects in academia.

As Christmas approached, I sat at my desk in the Google office and thought about that Christmas at Kinder's house. I was broke, but I was still writing and still able to support myself and Gracie, barely. I replayed all the missteps that had led me to the MFA at Chatham and eventually to my job at HCL. I sat at my desk and thought about Chuck and Diane and my daughter while our operations manager advised us on ways to "hack" the TVC system by taking advantage of full-time employees who didn't want to wrap their own gifts. Management also recommended we drive for Uber or take other jobs on Google's version of MTurk for extra cash.

I vacillated between despair and optimism, which was at least a step up from unrelenting self-hatred. I had so much downtime at work, I was getting really good at pinball. I realized the main reason for the downtime was the way they delegated our tasks. Essentially, my old job at Thermo Fisher was broken into ten steps, and each individual step was given to a different team on Google Shopping. This allowed HCL to devalue our labor, and it made workers easy to replace. I didn't need to know how the whole back end of our e-commerce system worked or how the taxonomy and labeling worked; I only needed to know if a product matched the others on the page. This tactic of using piecemeal work has been used by a lot of corporations in a lot of different sectors. It's one reason why plants use assembly lines. Seventy-five percent of the work I did was a matching game. The interesting ad hoc work was getting scarce. Turnover was high.

I had written a few pages and finished my task work for the day, so I decided to play pinball. I took a different route upstairs so I could get to know more of the building and so my supervisor wouldn't see me heading to the game room every day at the same time after lunch. I still hadn't found a job, despite dozens of interviews. I felt alone and sad most of the time, but thinking about getting through some rough Christmases with the help of others gave me something to draw from at least.

I was always hesitant to ask for help. I don't know to what extent this was conditioned in me, but it took a concerted effort to let someone know I was struggling. It didn't make sense because I rarely felt burdened when someone asked me for help. I was almost always appreciative that I could do something that would take my mind off of my own problems. With this in mind, as I ran through the prompts of the *Terminator* pinball machine, I thought more about unionizing. The more I thought about it, the more it seemed like the only way to improve things at work. Even if it failed, it would be worth the effort. I should at least ask someone how to go about it.

A few days after this, I mentioned it to May, and she said she might know someone who could help.

February 2019
DAMON

I had a few comically bad interviews at the start of the new year. I got so nervous that I forgot the most basic elements of the positions and the companies where I was applying. But every once in a while, I'd nail it. Then I'd get a second interview, submit a tech writing sample, maybe get a third interview, and then be rejected. With one potential job, this happened on a Tuesday in late January, and after the rejection, I spiraled hard. I couldn't sleep, and I lost my appetite. I used prorated future PTO and called off the rest of the week. I got Gracie to the bus stop and then slept on the couch for most of the day. I didn't see a lot of daylight. I was doing the bare minimum as a parent. I could feel myself becoming more withdrawn, but I also felt like I couldn't do anything to stop it. The optimism of Christmas time was gone. I was seriously considering quitting and living off my tax return for a couple months and just hoping for the best. I was ready to bus tables again if it meant getting out of HCL.

That weekend, May gave me the number of a United Steelworkers (USW) organizer named Damon, who she knew through a mutual friend. At work, I talked to George in the laundry room, and he said he would support an organizing effort. I had no idea how to start a union, but I figured if George was on board, we had a shot. I didn't feel like it was an insane idea because he thought it could work too. When professional organizers say it only takes two people talking about work grievances to start a union, it literally did in our case. Luckily for us, George was one of those people. He was a natural leader, and everyone at work liked him.

A couple weeks after I talked to George, I texted Damon. He was working on the faculty campaign at Pitt, and he agreed to meet for coffee at a Starbucks near campus to see if USW could help us out. Worst-case scenario, I thought, I would learn something about the basics of unionizing. And if May knew this guy, he was solid. I trusted her more than anyone, and she was supportive and understanding while I was struggling to find a new job.

On the day we were scheduled to meet, I circled the block and looked for a parking space, feeling guilty for taking time off work and not spending it with my kid. At night, it was difficult for me do too much more than ask Gracie cursory questions about her homework, make a quick dinner, and maybe watch TV with her. She had always been reserved, but she'd grown even quieter since her mom had died. We were both looking forward to softball in the spring, but February in Pittsburgh felt like it went on forever. I thought about parenting and work, and time and money, while I scanned the side streets off Forbes Avenue. It was nothing but fire hydrants and yellow curbs everywhere. I parked at a meter by the library and walked toward Posvar Hall, where Forbes Field used to sit. Oakland smelled like fryer grease and rain.

I had worked at this Starbucks for a few years while I was in college. One of my coworkers went on to become a renowned artist. I remembered him and other old friends and old jobs as I headed toward the main drag on Forbes. The nostalgia of being in Oakland was worse in the fall when I had time to loiter, the cool air and the dying summer signifying a new school year and a fresh start. It was too cold now to do more than jolt loose a few memories.

It felt like I was always starting over. There was always some new, arbitrary demarcation I could find to use as a reference point. Birthdays, jobs, sobriety dates, new baseball seasons: with each one, I told myself that this would be the time I finally figured out how to manage my time and live a balanced life. I had started over when I'd moved to Pittsburgh as a twenty-year-old. Almost two decades later, I thought I would have more answers.

Traffic picked up as the sun sank behind the afternoon clouds. I stood on the corner of Forbes and Atwood, trying to finish my cigarette before my fingers went numb. Across the street, there had once been a music venue and a head shop on the second story of a used CD store. Now there was a cell phone store, a burrito chain, and a bank. It felt like the temperature had dropped ten degrees.

I thought about switching careers, but then I realized I didn't have a career to switch from. In the past, if I'd had enough time to write, that felt liberating. But now, I couldn't think, let alone engage seriously with words on the page. If I wasn't going to make enough money to live comfortably, I needed time to write and time to spend with my child. I could barely pull myself out of bed in the morning. Balancing a day job, a side job, and a dead-before-it-started literary career as a single parent was near impossible. I wasn't doing any of these things well. I struggled and worked sixty hours a week. I was nearing forty, loaded with debt and lacking savings. All I did was work and worry. I felt buried.

I pulled one hand out of its glove, lit another cigarette, and stood by the window that ran down Atwood Street, where I used to take breaks fifteen years ago. My cold breath and clouds of cigarette smoke billowed upward, and it felt like 2003 again until I looked back across the street and saw that all the places I used to visit were gone. Everything felt more congested than it had between 2002 and 2006, when I barhopped and couch surfed my way through undergrad.

I didn't know what a meeting with a union organizer would accomplish, but I was out of ideas. I hadn't felt this beaten up since I'd first gotten sober, and my job was at the root of it. I brought years of baggage into my first meeting with Damon. We got our coffee, left the crowded shop, and walked up Forbes to the Pitt Law building, where there was a quiet place to sit and talk in the lobby.

We sat down at a wobbly high-top table. It was just a cylinder jutting out of the ground, and I had to half-straddle it to get close enough to set my coffee cup down.

"Tell me about work," Damon said. "Other than money, what are your biggest issues?"

I told him about having to use PTO when the office wasn't open, and he wrote that down. I told him about our lack of sick days, about how the recruiter's initial description of the job didn't match up with what it was like, and about the pay disparity between TVCs and full-time employees. Damon was attentive as we talked for about a half hour.

"What do you think?" I asked.

"I'm not sure. I'll ask my boss about this and get back to you. In the meantime, try to get one or two other folks who are interested, and see if they'd like to meet for a beer or coffee."

"I can do that," I said. "I have only talked to one other person, but they are on board."

We shook hands and parted ways. I doubted I'd ever see him again.

Back outside, I lit a cigarette and looked toward downtown. Way down Forbes, near the CVS, stood Magee-Womens Hospital where Gracie was born in 2007—like I'd just walked in a twelve-year circle. By the time I got home, I was drained and overwhelmed. I ordered takeout. I worried about the expense, but Gracie was happy. We sat and ate pizza in front of the TV and talked about our days.

March 2019
POST-SURGERY, WATCHING THE NEIGHBORS

I swallowed my last painkiller and rolled over on the couch. I lay on my side and watched TV. I was near sleep when the neighbors across the street started screaming at each other.

I picked myself up off the couch, limped to the window, and watched.

They yelled about money.

The guy threw her stuff out on the street.

She cursed at him while she gathered her clothes.

Then he went back inside, and she waited on the curb until someone in a black truck picked her up.

Their kids seemed happy whenever I saw them on the street. I'd only said hello in passing. When I saw them walking around or playing with their kids in the big parking lot on the corner, I wanted to pull one of the parents aside and talk to them. Maybe I could help. Maybe one day I'd tell them what I'd been through with Gracie's mom, how I knew there were only a few ways the situation could end and that most of them were bad. But I didn't. Even if I'd had the nerve, it wouldn't do any good. If someone had offered Jane help when we were in that situation, she would have just tried to rob them. I assume this because she robbed me when I tried to help her.

I emptied the piss bag and checked on the couple again. There was enough of a gap under their blinds across the street that I could see movement and dim light in their upstairs room. There was a two-by-four sitting diagonally in one of the windows. It was quiet again. I shuffled downstairs and outside to smoke. Across the street, a beat-up car with Ohio plates sat in the driveway. The dome light was on, and the mom was in the front seat, getting high. The dome light went off. I stayed on the stoop long enough to finish my cigarette and get back inside before they got out of the car and saw me there. I thought, if I was a better neighbor, if I really cared about my community and the effects of the opioid crisis as much as I thought I did, I would have figured out a way to help. But I just looked the other way and crossed my fingers for them. There was Narcan in the mutual aid stand up the street. I suppose I could have dropped some Prevention Point literature in their mailbox, but I mostly hoped they'd move away and live happily ever after.

When Jane's grandfather was dying in Shaler, she stole his cancer medication. She bought thousands of dollars of liquor on her grandmother's credit card and either sold it or traded it for dope. Some of the booze was for Susan, JR's mom. Eventually, Jane wore out her welcome at her grandparents' place, and she and Gracie moved back to Evans City, where they stayed until Jane's next overdose.

For months after they'd moved, Gracie asked me about JR and when we could go see him again. I tried to call Susan, but she never answered. JR's dad was always away on business, but Susan was pleasant enough. Mostly, she was glad that Gracie would go over there and play *Minecraft* or whatever make-believe backyard game she and JR would come up with. Jane, high on dope, would sit out back with Susan on the patio, just two moms hanging out, watching their kids. Then Jane would leave for a bit and score dope and come back. She'd bring booze to sell Susan, who would switch from wine to a cocktail after dinner. Gracie and JR played in that way you can when you're young and the days feel endless but not hopeless.

I thought if the laws were different, if Jane could have gotten a job even though she was a felon, if there were safe injection sites and decriminalized drugs, if addiction didn't carry such a stigma, she might have had a chance. She waited tables until she couldn't, and then she found a way to support her habit without a job. Unable to find community in recovery or in the normal world, she kept getting high. She loved Gracie more than anything, and she still got high until it killed her.

For almost ten years, I'd endured meltdowns and public shouting matches and unreliable pick-ups and drop-offs on holidays. I dealt with increasingly erratic behavior and gaslighting. I spent tens of thousands of dollars in family court. It pushed the limits of my sympathy for addicts.

———————

When my neighbors fought out in the street, I watched. After two weeks tethered to my couch, I'd learned their routine. In my mind, the dad had a second-shift job and a babysitter nearby, because I always saw him walking with the kids toward Liberty Avenue around 3:00 p.m. and then coming home again around one in the morning. The mom was transient. She'd come around regularly on weekends. I wondered if the family across the street had watched me and Jane fight on the sidewalk or heard us yelling at each other while Gracie waited upstairs in my apartment. Surely there were neighbors in Shaler who had heard us yelling at each other and watched Jane chase after my car. I wondered if those neighbors felt lucky that they weren't in a similar situation, or if they were appalled at us—disgusted by the whole messy situation that popped up in their quiet development every so often. Maybe they made up backstories too. I wondered who they blamed it on. What did those neighbors think when Jane pulled Gracie away from me—out of my arms—and ran back into her grandparents' house? What did they think when she called the cops on me repeatedly, for just showing up on my visitation days? They couldn't know that I just wanted her to follow the custody order. Maybe no one saw anything, or, if they did, maybe they didn't give it a second thought.

I was making up stories about people making up stories about me. The pills were going to my head. I contemplated some kind of *Rear Window* situation, playing through many hypothetical instances of seeing the neighbors dispose of a body—maybe someone who overdosed—or being a nosy neighbor who threatened to call the police or Child, Youth, and Family Services. The cops wouldn't ever believe some unreliable witness who was on all these painkillers. At the end, I would die or go crazy.

Their apartment was owned by a guy who lived in Utah. He'd bought it sight unseen and hired a property manager to take care of it. He was out in Park City or someplace like that, collecting rent while his property was continually on the edge of a domestic dispute or an overdose case. I was madder that the house was owned by an absentee landlord than the fact that it was occupied by a junkie. I don't think the two were really related. Mostly, I just hoped the kids never woke up and found their mother's body. I hoped everyone was OK.

The next morning, I left my follow-up appointment in Wexford with a prescription for non-opioid pain meds and orders to stay home for another week. The catheter was gone, and so were the stitches. On my way back to the city, I got a call from the short-term disability office asking for a new doctor's note to extend my payments. My doctor had left a message with them, but he also needed to fax over a letter.

"You want me to go back to work? My ass is still bleeding. The doctor said to take another week."

"I need a letter."

"They sent it."

"I haven't received it."

"I'm driving. I'll email you when I get home."

When I got home, I called the doctor's office to make sure they had sent a letter. Then I called the short-term disability office to make sure they had received it. I emailed the doctor and the disability office so I would have a paper trail showing I had done my due diligence. Donald told me I could come back midweek despite the leave-without-pay situation. He said it didn't apply to days off for disability.

Spring break was over. Gracie was back in school. I no longer had to wear adult diapers, and even though my ass was still bleeding and sore, I spent five days writing. It was the best week of the year. I applied to more jobs. I felt good, mentally, for the first time since I had started at Google. I still used pain meds to help me sleep. When Gracie came home from school, I helped her with her homework. After dinner, I lay on the couch and watched *The Sarah Connor Chronicles*. I was midway through the second season when I got a text.

"Ben, it's Damon from USW. Another worker at HCL reached out about unionizing. Can we all meet for coffee next week?"

I texted George and told him the news. There were three of us now.

PART II

March 2019
HOW TO FORM A UNION

Step 1: Talk to Coworkers and Form an Organizing Committee
Step 2: Map and Assess
Step 3: Circulate Cards and Inoculate Coworkers
Step 4: File Signed Cards with NLRB for Election
Step 5: Continue to Assess, Strengthen Base, and Inoculate against Union-Busting Campaign
Step 6: Vote

If Election is Won:
Step 7: Nominate and Elect Bargaining Committee
Step 8: Create Bargaining Survey for Bargaining Unit with Contract Priorities
Step 9: Negotiate Contract
Step 10: Bring Contract to Bargaining Unit for a Vote

If Contract is Ratified:
Step 11: Elect Officers

March 2019
OUR FIRST MEETING

George and Damon were already at Zeke's in East Liberty when I walked in. It smelled like coffee and cleaning products. One of the baristas was mopping the floor behind the counter. I set my notebook down on the table and took my place in line. I wondered who the other person was who had reached out to USW. After I sat down with my coffee, Carl, a quiet guy who I'd hardly spoken to, walked in and sat with us.

Carl had been with the company for five years. He was one of the first new hires after HCL took over the Google Shopping contract, and his salary had started at $35,000. After five years, he was still making less than I was, and I'd started six months ago. Carl was on Catalogs, too, but he'd started on a different team that did consistently complex work. He spoke Japanese. He was eleven years younger than I was and more laid back. He had community organizing experience and had wanted to start a union years ago.

I drank coffee and took notes while George and Carl talked to Damon about work. They told him about the low pay, random starting salaries, and opaque policies. They had more insight into the workplace than I did. They said the structure of the office benefited those in charge. Bosses worked from home. Bosses promoted their friends and talked down to everyone else while praising Google and the office perks they provided.

"We have to use PTO when the office is closed, and we don't get sick days," Carl said, repeating a common complaint.

Both George and Carl mentioned the odd policy of having to use unpaid leave in week-long increments, even when it was combined with PTO. Damon asked if there was an employee handbook, but at that time, none of us knew if one existed. The only policy information we could find was an internal wiki site on the HCL at Google webpage, and not everyone followed it. Rules varied from team to team depending on the supervisor. Some workers were granted extra weeks of vacation instead of raises. Some employees were forced to work unpaid overtime and take extra leave the following week instead of getting paid time and a half. Workers on some teams couldn't work more than nine hours a day, while some of us worked two eleven-hour days, a ten-hour day, and two four-hour days. Some people were allowed to use unpaid time off in hour increments, while Nancy had insisted we needed to take entire weeks off. Some teams were not allowed to leave their desks for more than two hours a day.

We wanted consistency and flexibility. We worked right next to people who made six-figure salaries and had unlimited sick time and three weeks of vacation. They had better health insurance, stock options, and they could work from home if they needed to. Full-time employees could afford daycare. They made so much that they didn't need all the office perks and benefits they received.

Damon listened and took notes. He explained the steps involved in forming a union. Because we were full-time, permanent employees with HCL, we could form a traditional union under the National Labor Relations Act (NLRA). He reiterated how important it was that we be the ones talking to our coworkers, not USW staffers. He was there to help, but it was our show. We had to grow our committee from the three of us to include at least nine people. The organizing committee needed to represent the whole workforce demographically and across different teams.

"You want everyone's perspective," Damon said. "Ideally, you'll get about 10 percent of the workforce. Recruit people who are well respected and good workers. Along those lines, make sure you all are ideal employees from here on out."

We were cautioned not to mention the union unless a recruit was completely on board.

"Go for the low-hanging fruit first," Damon said. "And if anyone says 'yes,' ask if they can introduce you to someone else they think would be interested."

We agreed to reconvene in a month and assess our progress.

March 2019
SECOND SOFTBALL SEASON

After the meeting with Damon and George, I went to the Boys and Girls Club to pick up Gracie. When I got there, she was outside on the softball field, practicing with the Ninth Ward's twelve-and-under team. For some reason, they played in a different rec league than the Shadyside team I helped coach. I walked over and sat on the bleachers and watched infield practice. My neck ached and my hands hurt. I was so tired. I worried that I wasn't sufficiently hiding how hard things were and that I was putting stress on Gracie too. As a kid, I never saw my parents struggle. My dad had a good job and was always there for me and my sisters. At the time, I didn't really appreciate it. I wondered what Gracie would think of me when she was in her thirties, navigating a rigged economy and the increasing effects of climate change. I hoped she would have an easier time than I'd had. I sat there, wishing I could do more to make it easier for her.

The team took a water break before batting practice. Gracie ran in from the outfield.

"Dad," she said, "Can I play on Ninth Ward too this summer?"

"Sure," I said. "But Shadyside games are the priority if they're on the same day because I have to coach."

"Can we go to Wendy's after practice?"

"No."

Gracie slumped her shoulders and rolled her eyes in a how-can-you-deny-me-a-Frosty look. She jogged back out to the field.

The Ninth Ward team played their home games in Lawrenceville, five minutes down the street from our apartment, and I didn't have to help coach. It got Gracie out of the house and around another group of kids. Anything to take her mind off of her mom for a couple hours was welcomed. While the girls hit neon yellow softballs in high arcs toward the outfield, I thought about the summers I had spent as a kid on baseball fields in New Jersey. Sometimes I wonder if the baseball part is incidental to the nostalgia or if it's the focal point. Maybe I just miss being young and not having responsibilities, and baseball was just what I did as a kid. Maybe it could have just as easily been another activity. But I loved playing baseball growing up. I went to games early to take extra batting practice. My dad hit me fly balls on his days off. If he wasn't coaching my team, he attended my games. Baseball also presented me with an aspect of parenting I could model successfully. It felt comfortable and familiar. Gracie's practice didn't end until the sun went down and the outfield lights flickered on.

April 2019
TECH ORGANIZING AND CHALLENGES

With traditional manufacturing jobs on the decline, USW had started organizing nonindustrial workplaces, such as the Carnegie Library and Museums and the faculty, staff, and grad students at Pitt. Around the time we were starting our union drive, a number of smaller nonprofits had also organized with USW. If we were successful, we would be charter members of the new tech organization USW had in the works. This was before the Communications Workers of America launched their Campaign to Organize Digital Employees, which benefited from a team of dedicated, full-time tech organizers as well as the great acronym CODE.

That spring, Kickstarter was in the news after workers were fired for organizing a union with the Office and Professional Employees International Union. Their campaign, one of the first attempts to organize full-time tech workers, was inspiring. The gaming industry was fighting against horrific working conditions. Content moderators and data center workers were organizing. So were digital media workers and contractors at social media companies. Uber and Lyft drivers in California were fighting for classification as full-time employees. For many of the big tech companies, the shine was beginning to wear off.

In addition to pay and benefits, full-time tech workers across the country organized around other workplace issues, including sexual harassment, discriminatory hiring practices, and forced arbitration. The second-class status of contractors had been an issue for decades in tech and other white-collar industries. Contractors in tech faced all the problems full-time employees did but they didn't enjoy the job security, benefits, or relatively high pay.

In California, Proposition 22, a ballot measure largely written by Uber, Lyft, DoorDash, and Instacart, was designed to let those companies evade labor law that would classify their drivers as employees. According to *Vice*, Uber spent over $200 million to help the ballot measure pass, keeping their drivers classified as independent contractors, which meant they were ineligible for benefits and unable to engage in collective bargaining. However, the National Employment Law Project reported, "The Alameda Superior Court of California recently ruled that Proposition 22 violates the California constitution and must be struck down in its entirety." The companies are going to fight this ruling.

Although the tech sector has seen an upswing in labor activism, tech companies have essentially unlimited resources to suppress dissent. They write their own laws, buy politicians, and funnel money into antiworker groups like the Chamber of Progress and the Coalition for Workforce Innovation (CWI).

The Chamber of Progress is a lobbying firm that claims to champion progressive policies for tech companies while actually suppressing legislation that would, among other things, reclassify gig workers as official employees or hold social media platforms accountable for spreading hate speech and misinformation. It is funded by tech companies and run by a former Google lobbyist. CWI is even scarier. The big five tech companies (Alphabet, Amazon, Meta, Apple, and Microsoft) are all represented in this coalition via another lobbyist group called TechNet. This group also includes gig companies and companies like Walmart, Amway, the Direct Selling Association, which is a "trade association" dedicated to multilevel marketing, and big union-busting law firms like Littler Mendelson. This group wants to end traditional employment. They want to make Proposition 22 national law, which would crush worker protections and job security across the country. The laws these lobbyists want have bipartisan support.

Lobbyist groups like these use left-sounding language in their press releases and communications. The language they use on their websites, which often touts the flexibility of independent work, frames their goals as pro-labor.

The big five tech companies are the most powerful and richest companies in the world. Without a revitalized labor movement, we are headed toward a future of employee-less businesses ruled by lawyers, real estate speculators, and do-nothing middlemen. Without unions, the rest of us will wind up in Amazon company towns propped up by cryptocurrency, doing piecemeal work, driving for Uber, and crowdfunding our health care.

April 2019
BUILDING THE COMMITTEE

George, Carl, and I began to meet weekly for lunch and to compare notes. Union avoidance firms advise managers and supervisors to look for employees who are spending time with different people as one of the signs they might be trying to organize. Because Carl, George, and I were all on the same team, it wasn't unusual for us to be seen together. But we still didn't take any chances. Whenever we met, we waited until the end of lunch service, when our bosses were back at their desks, and we each took a different route to the cafeteria.

Over the course of a month, we had one-on-one conversations with potential recruits for the committee. We kept quiet and never used our work computers for anything related to the union.

I had no idea what I was doing, but I read as much as I could about organizing and tried to take Damon's advice. Listening was the most important thing. I listened to my coworkers and related my work experiences to theirs.

I talked to the people I knew, and to the smokers, many of whom were also pissed about the reality of the job versus how it was pitched to them. I heard a common refrain: the recruiters had misrepresented the job drastically. They hyped up the free lunches and the possibility of getting a job with Google; they were not forthcoming about the raises, and they lied about increases in PTO, the possibility of remote work, and the promotions. They said what they needed to get people in the door. The honeymoon period for this job was laughably short, even with all the perks. This was how it was when I worked at other jobs too. They'd burn out anyone who made an effort and promote undeserving people who had friends in management. Soon, others would realize they had no prospects of advancement, and they would quit and be replaced with cheaper new hires.

I started to get to know my coworkers better as I talked to them, but I still wasn't really close to anyone. Early on, Carl, and especially George, did the heavy lifting. By the end of April, we had recruited more than 10 percent of the workforce to join the committee—nine of the ten teams were represented. All we needed was a meeting place.

May 2019
MAPPING

The USW building had a classic feel to it. Its architecture and design nodded toward labor's better days and the union's rich history in Pittsburgh. The windows were diamond-shaped, and there was a hint of cigarette smoke baked into the walls. When I met George and Carl there, it was after 6:00 p.m., and the place was mostly empty. We took the elevator to the seventh floor and sat at a big table in a common area with a whiteboard and some filing cabinets. Labor posters from different eras hung on the doors. Magazines about organizing and labor rights were piled up on the shelves. A bag of pretzels and some baked goods sat on a row of filing cabinets. Stale coffee coagulated in the pot. I filled a cup with water, and the deep glug of the cooler calmed me as the bubbles rose to the top. I was amazed we'd made it this far. I could not believe we were discussing strategy in the Steelworkers building.

Maria, USW's head of organizing, gave a brief but rousing speech about unions, the Steelworkers, and our bravery in standing up to management.

We outlined the L-shaped floor plan of the Google office on the whiteboard, marking where the bosses sat, the blind spots the cameras couldn't see, and all the entrances and exits. We noted all the break rooms, game rooms, microkitchens, and patios. George brought employee lists that included everyone's title and team, and we used it to fill in our spreadsheet and our map, noting team leads, lead analysts, and trainers. Our spreadsheet had columns for everyone's team, their most pressing workplace issues, and their contact info if they had given the OK for us to use it. Eventually, we ranked everyone on a scale of one to four: Activist (one), Supporter (two), No Opinion (three), and Anti-Union (four). We also included a column for the date someone on the committee had last talked to them and how the conversation went.

We made notes of what time our bosses and supervisors came and went, what exits and entrances they used, where and when they took breaks, and where they parked. It was dark out by the time we finished. We left downtown tired but confident.

May 2019
THE FIRST COMMITTEE MEETING

In those first few months, I felt like I wasn't doing enough. Except for George, no one I had asked wanted to be on the committee. But I found a place to hold our meetings, and I was happy to contribute something concrete. May put me in touch with a friend of hers who ran a tech crafting space in a repurposed theater between North Oakland and Polish Hill. It was a quiet, comfortable spot filled with intersectional feminist art and high-tech gadgetry. It was far enough from the office that no one would see us gathering there but close enough that we could stop by after work without too much hassle. I got there early to set up.

My experience in Alcoholics Anonymous was coming in handy: I could listen to people's problems, keep a secret, and perform menial tasks as a way to be of service to the group. I pushed tables together, arranged chairs in a circle, and wondered who we'd recruited. Then I sat and waited while everyone else trickled in.

A few committee members had previous organizing and union experience. Rachel was in the American Postal Workers union before coming to HCL, George had been in a teachers' union, and Delia was a former NYU graduate union organizer and had gone on strike in the early 2000s. I recognized Paige and Sarah from the couches on the balcony overlooking Penn Avenue. I often passed them when I was going to work outside in the afternoons.

As we got to work, the union started to feel less abstract. We were about to take a big risk together. That first evening, we introduced ourselves and went over some basic organizing training with Damon. We covered frequently asked questions and common misconceptions. Dues were only 1.45 percent of our pay; half of that stayed with the local to use how we saw fit, and the other half went toward lawyers and the USW strike fund. No one paid dues until we ratified a contract, and we wouldn't ratify a contract that didn't include wage increases that would more than cover our dues. All decisions were made democratically. The international gave us resources, but how we ran things was ultimately up to us. We were the union.

Everyone on the committee exchanged numbers and started a group chat on an encrypted app. We shared permissions to the spreadsheet, and the new committee members added what they could. Even taking care of these basic housekeeping items felt great. It was weird seeing everyone in a different context, but there was a sense of excitement. We all had a secret, but it was more than that. There was an optimism that we could do something concrete to help each other. After that first meeting, I was ready to fight for my coworkers, and I knew the people in the room would fight for me too.

We met every week after that. Each meeting brought more good news. We kept quiet, and we made sure to reiterate the importance of discretion to everyone we talked to. Our numbers were good. We needed at least half of the workforce on our side before we circulated union cards, at which point management would probably find out. We were on pace to get there in a matter of weeks.

Damon chaired the committee meetings. George held us together when we lost momentum or started to drift. He was the unquestioned leader in the office and of our committee. We all felt hopeful that we could change the culture at work and get long overdue raises and basic improvements. It was inspiring.

The isolation I'd felt for months at the job began to dissipate. Our committee grew. The next wave of recruits brought in Sam, Leah, and Will. Sam and I were happy to see each other again, and we recounted our initial meeting in the seventh-floor waiting room. Sam had worked with leftist organizations as a community organizer in the past. They had posted a joke in the work chat that we needed a union, and Carl had approached them later that day. Leah was more reserved. She was a former professor with a doctorate in the humanities and had been laid off and rehired by HCL. We had mutual friends outside work.

Will was from Los Angeles and had worked as a Google contractor out there before transferring to Pittsburgh. Everyone thought he was a manager at first. He had created a new team, Falcon, along with all its processes, and had been unofficially running it. He said he was denied an official promotion to team lead because he made too much money already; they wouldn't give him a raise, just added responsibility. He didn't have direct reports, though, so he fit the National Labor Relations Act's criteria of a bargaining unit member. His enthusiasm for the union was fueled primarily by his hatred of management. He worked closely with his team lead, so he had access to information the rest of us didn't. He told us to expect a bunch of new hires in the near future as management filled the roles for the team he'd created. He talked at great length about how much he despised the bosses, often ranting about them for so long that Damon had to redirect the conversation.

Damon reminded us that we had to follow labor law much more closely than our bosses did. We had to be perfect, operating within the very narrow set of parameters set up by the NLRA. Any misstep would give management an opening to decertify a potential election. The most important thing at this stage was to keep management in the dark for as long as possible.

May 2019
SCHOOL

One Friday around lunchtime, I got an email from Gracie's school, saying that she had three Ds halfway through the marking period and would need to complete a large backlog of assignments to get her grades up by the end of the year. I felt horrible. I'd been so focused on the union and work that I'd completely ignored her grades. She'd always been on the honor roll in elementary school. The year before, she had slipped a bit after her mother's death, but I gave her a pass. This year, though, she fell behind and was running out of time to get caught up.

I snuck out of work early, went home, and cleaned her room. I scraped melted candy wrappers off the floor behind the radiator and swept up empty chip bags. I gathered and sorted piles of dirty clothes and put away her toys, most of which she hadn't played with in years. Her bed was still full of stuffed animals. I went grocery shopping, got dinner ready, and then picked her up from the Boys and Girls Club.

"You need to do the work," I said.

"Sorry, Dad."

After dinner, we sat down on the couch in the living room and got ready to go over all the homework she was missing. Pittsburgh Public Schools used two different online platforms, and some of the teachers used their own online system. I had to find the letter containing my login ID and password to sign into the system so I could see her grades and assignments that were spread across these disparate locations. The syllabus also referenced an online math program they used that I didn't have access to.

"Sorry," she said. "I can't keep track of everything."

"When I asked if your homework was done, you said, 'yes.'"

"I'm sorry."

"I'll help you, but you have to tell me you need help."

We sat together and worked through her problems. She had a bunch of creative writing assignments due that we'd have to finish up that weekend. We went subject by subject, and we almost got caught up.

Gracie had also been late a few times the previous year and had missed a handful of days. I'd let her stay home if she said she wasn't up for it. I didn't think it helped anyone to be overly hard on her after everything she'd been through. I was searching for a therapist for her, but it took forever to find someone who accepted my insurance and worked with kids. UPMC put us on a waiting list, but they said it would be months before she could see someone.

When we were almost finished, I said, "Listen, you won't be able to do this in high school, and if you want to go to college, you're going to need to get good grades and get scholarships. I can't afford to pay for it." I sat up and rubbed my eyes. I regretted being so blunt.

"I know."

"You're smart. Just do the work. Do as much of it as you can in class."

"I do."

She turned off her laptop and we played the new Mario Bros. game I'd bought her for Christmas. For the next couple of hours, we jumped and fire-balled our way through a few levels while Gracie played pop music on her phone. It was as much quality time as we got.

"Is everything OK?" I asked.

"I'm tired of people calling me brave," she said. "I don't want to be brave. I want a mom."

She spoke very matter-of-factly.

"I know," I said. "It's not fair."

"I remember Mom falling a lot and going to all these weird people's houses. It's messed up. I don't know why she did it."

"She thought she had to. When you're addicted to drugs, you do things you wouldn't normally."

"Why'd she do heroin in the first place?"

"I don't know," I said.

Jane once told me about the first time she shot dope. The way she described it, it was just another random night with her boyfriend after a Bob Weir show. But after that night, she said, heroin was all that mattered. Nothing else would do it for her. She was barely out of high school.

I thought about the randomness of her addiction and my own sobriety—grateful I'd never tried heroin—while Gracie and I warped to a water level. We swam around in penguin suits and shot ice balls at pixelated fish and squid.

"Your mom loved you. She was so funny when she wasn't on drugs."

"Did you break up with her because of drugs?"

"Basically," I said.

Really, the relationship had just run its course and I wanted to leave. Then it got complicated when Jane got pregnant and Gracie was born. It got even more so after Jane relapsed.

Gracie sat quietly and didn't respond. We played a little while longer and then it was time for her to go to bed. I stayed up late watching TV, worrying about her and unsure of how to help.

May 2019
UNDER THE LIGHTS

Gracie's first game with her new Ninth Ward team was a doubleheader way down in Oakdale. We made the opening day trip thirty-five minutes west on an unseasonably cold morning in late spring.

"I wish I could play fast-pitch," Gracie said.

"That would be fun," I said.

For some reason, there wasn't recreational fast-pitch softball for kids in the city of Pittsburgh. All I found online were travel teams in the suburbs. Those cost thousands of dollars and would require thousands more for gas and lodging costs. Gracie wasn't a prodigy, but she was a good hitter and she enjoyed playing. Mostly, she liked the social aspect of it, but it was fun to see her develop a competitive streak too.

"I like this team more," she said. "We're better than Shadyside."

"Most of the girls have been playing longer," I said.

We got to the game. The field was at the bottom of a big hill with a path that wound around to the dugouts and snack bar. I sat behind the fence in right field and watched Gracie's team lose the first game by almost twenty runs and the second game by ten. They were overmatched, but at least they kept it close for a few innings in the second game. I wondered why Oakdale was on the schedule, and I was glad we wouldn't have to make the long trip again. Gracie played pretty well and wasn't too upset on the ride home.

"They hit the ball so far," she said.

"They must all be twelve. Your team is younger."

We sat in silence for a bit. Rides home were more fun after victories, when we could recount all her team's hits and good plays instead of how many times the other team had hit a home run. We stopped at Wendy's on our way back to the city.

That summer, we spent most of our nights under the lights at softball fields across Allegheny County. Gracie had softball practice or a game five nights a week. The team I coached was not very competitive, but it was still fun. Going to Ninth Ward games and watching was more relaxing, though, and they won more than half their games. Gracie also played up a division when the Ninth Ward's fifteen-and-under team was short players. This was due mostly to us living close to the field and being available on short notice, but she held her own as an eleven-year-old when she moved up.

When Gracie didn't have softball games, I played baseball, and once school was over, she'd come watch. After games, she'd tell me the same things about my swing that I told her.

"You need to stay back, Dad."

"I know, honey."

I've been trying to develop some late-career power and mostly failing. I still got on base enough to be an effective bottom of the order hitter in our low-level adult league. In a short season, though, it's possible to do everything right and still not put up good numbers if the ball doesn't find the gaps. But it usually evens out. Working on the process—pitch recognition, staying on plane, and making better contact—is still worthwhile to me, even if it mostly results in singles. That summer, my team had another good regular season and an early playoff exit. I'm still waiting for our luck to even out in the postseason so I can win one more championship with my friends before I quit.

The satisfaction of hitting a ball on the barrel with a wood bat in my late thirties is amazing, but nothing compared to watching Gracie do well in a game. I was proud, sure, but I was also happy that she was happy. Happiness had been in such short supply for both of us.

Her first double of the season came in a night game against Reserve Township. She hit a line drive down the left field line and cleared the bases. They lost the game, but it was close, and I could see all the extra practice was paying off. She became a dependable hitter and moved up in the batting order. On a dry day at the field in Lawrenceville, which didn't have a fence, she put one in between the outfielders and ran around the bases for a home run. We'd stop for fast food after away games and talk about how she did and how the team looked overall. We'd talk about hitting, one of my favorite things. It was precious time together, and we grew closer.

Those moments were fleeting, though, and I knew it. When high school started and she became more independent, we would have less time like this together. I still regretted not appreciating playing baseball games that mattered and the summer days I spent with my dad when I was a teenager. I had mostly forgiven myself for being a stupid kid who didn't realize how good he had it, but sometimes the regret turned into deep remorse. I didn't want to have similar regrets about not spending time with my daughter. I put a lot of pressure on myself to be a good dad. The pressure grew in 2016 after Jane relapsed and I was awarded primary custody; it grew when I was awarded sole custody in 2017 and especially after Jane died. The weight I felt I needed to carry lessened during the summer months. I don't know if I appreciated that time more because Jane had died and I felt it was even more important to be there for my kid. Maybe I felt too self-satisfied with just doing normal dad stuff, clearing a very low bar by very little.

I was constantly overthinking things, but I knew Gracie was happy when she hit a double or made a catch at third base. I could see that joy. I equated that joy with me doing something right, or at least not getting in the way. Maybe my view of it all is self-centered, and I don't really matter that much, but I wanted her to know I was trying.

In the smaller moments, when the cloud of her mom's death wasn't as dark and didn't loom so heavily, we could just enjoy our time together and it felt great. I wasn't thinking about myself, I was thinking about Gracie and how much I loved her. I hoped she'd carry that with her.

June 2019
ASSESSMENT

My days at work were spent organizing the union drive and hedging my bets by applying to jobs. They were two high-stress endeavors, but I finally felt like I was doing something worthwhile during the hours I was stuck in the office building. When it seemed certain we were going to get to an election, I stopped applying to other jobs and started the first, very rough drafts of this book. I figured I'd stick around at least until the vote happened so I could see how the story ended.

By late June, we'd talked to over half the workforce and had gotten positive responses from almost everyone. We didn't make any promises other than the chance to negotiate. We would have cards to sign soon if there was enough support. We asked people what they wanted to change about their job, and we listened. We asked open-ended questions and empathized. We offered people a way to make things better. Whenever possible, we asked them to take some small action, like coming to a meeting or introducing the idea to someone else who might be interested.

We avoided a few people we thought were considered supervisors, and we still weren't sure whether trainers could be included in the bargaining unit. It was tough to ask people randomly if they had a role in interviewing, making subjective decisions about potential new hires, or if they had any role in hiring, firing, or assigning work—the other NLRB guidelines for what made someone a manager. There were also a couple well-respected folks who we didn't get to follow up with in time, but we talked to everyone at least once.

Union organizers are supposed to identify organic workplace leaders and get them involved and readied for postelection leadership positions. George was the lynchpin; he had a way of talking to people and empowering them without being cheesy about it, and he had previous experience with the teachers' union in New York. I wouldn't have reached out to Damon if George hadn't been on board, and we wouldn't have had almost all of Catalogs on board if not for him. He'd been at the company for years, was a great employee, and had a relatively positive attitude about all of it.

"I ask people if they want to go for a walk with my dog, and then I offer them the chance to take power in their workplace," George said at one meeting. "People like the idea of having a seat at the table. We've dealt with bullshit for years and people are tired of it."

One day, I ran into Mark and Walt out on the balcony. It was near the deadline to get everyone assessed prior to the card drop. I'd been agitating and commiserating with the two of them for months off and on, but I didn't feel good about broaching the subject of a union. I said we were trying to form a union as a solution to all the bullshit we were dealing with, and I asked if they wanted to help. They said it wasn't worth the effort, but they didn't dismiss it outright.

"Maybe when I was younger, I would have, but now, I don't know. What's the point?" Mark said. "I just want to do my work, go home, and drink beer in my garage."

He'd been working nights as a janitor but had quit because he hurt his back.

Walt spoke up.

"I think if we asked Nancy about it, it might be good," he said. "She seems really understanding."

"Telling Nancy would be a bad idea," I said. "Don't tell management, and if you want to get involved, let me know."

They agreed not to say anything, and I let it drop. I marked them as threes and made a note to have someone else approach them next time. Maybe if they saw that more people were involved, they'd get on board too. As tense as things would get between the organizers and the anti-union people, nobody said a word until it was too late for management to do anything about it.

June 2019
MANAGEMENT IS YOUR HR

Near the end of the second quarter, Jeffrey, the deputy general manager, showed his face in the building for the first time that year. All the HCL employees filed into the big conference room named after a Nabisco snack cracker. I sat close to the front with my teammates. I would normally skip this kind of meeting, but I was trying to be a good employee. We decided to ask questions.

Jeffrey and Amar, the head of HR, gave a presentation about how much money we were making for Google and HCL and why we should be proud of our efforts. They clicked through the slide deck on the huge screen set into the wall and took turns narrating.

Amar could not contain his enthusiasm. I almost felt bad for him.

"You have helped HCL achieve our most profitable year ever. We are close to $10 billion in revenue. Our Google account is especially lucrative, and we could not do it without you."

Workers who had been at the company for five years had seen their free parking taken away and their ability to take single days off eliminated. They had received, at most, a 4 percent increase in wages (0.75% annually) over that time. Meanwhile, the national cost of living had increased more than 8 percent between 2014 and 2019. And that didn't account for the disproportionate increase in Pittsburgh's housing costs.

After the presentation came the Q and A.

George asked what Jeffrey and Amar thought about the high turnover, the growing number of HCL's negative reviews on Glassdoor, and how they justified giving us 1 percent raises.

Why should we feel proud when the company didn't give any indication that it cared about us? Amar answered the question as if George was worried about the company's reputation. He told George not to worry and then admitted to paying people to put pro-HCL comments up on job sites. He walked around the podium, looked at George, and said, "It takes courage to stand up, but sometimes it takes even more courage to sit down."

George had asked a well-articulated, concise question about working conditions, and Amar had suggested he sit down. This got a rise from the crowd. An audible groan went up from the balcony.

Someone asked why we had to come to work during snow days and why we didn't get paid days off when the office was closed. I asked how much money HCL made from the Google contract and was ignored. Carl asked about wages. A few people looked startled by the confrontation. Others followed with more tough questions, all of which management dodged or ignored.

Eva asked why the holiday party had been canceled for the third straight year, despite the funds being allocated for it. When someone asked why we didn't get sick days, Amar couldn't believe we would ever miss a day of work.

George said, "Why is it so hard to get in touch with HR over workplace issues?"

Amar responded, "Management is your HR."

We were supposed to go to management with all our problems. We were told they had answers for us. The idea that we would have a problem with management was not addressed.

Management is your HR. Amar delivered this line like a televangelist. I think he honestly believed in the company, and he was far enough removed from our work experience that he couldn't believe we would have any real grievances. We should be proud of how our productivity translated into corporate profit. He was right, though. HR only cared about the interests of management, not the workers. But to not even pretend that we had someone to go to with problems with our supervisors was hilarious and sad.

The presentation ended. We all went back to work.

July 2019
MY LAST VICTORY AS A YOUTH LEAGUE SOFTBALL COACH

Gracie's team had a tough season overall. For most of the games, our goal was just to avoid the other team scoring six runs in an inning, which was the maximum allowed by the rules. We were the second-worst team in the league. But in the last regular season game, we had a chance. We were playing a team, the Tigers, we'd beat once earlier in the season. Their infield was immaculately manicured, and because it sat on what was once a football field, the outfield was cavernous in left and short in right. They were still winless but had improved over the course of the season. They knew this was their only chance at victory.

Throughout the regular season, we'd played everyone at every position. This game, we'd try to win. The only thing we would not do in pursuit of victory was tell the girls who couldn't hit not to swing. Maybe, if it was close, we'd have them take a strike, but looking for walks in slow-pitch was brutal.

Gracie and I discussed this on the way to the game.

"We can beat the Tigers," she said.

"I know. It should be a good game," I said. "I think we have everyone today."

"Am I going to have to play outfield?"

"Just play where the coach tells you. Outfield is important too. The best Pirates ever played outfield: Barry Bonds, Roberto Clemente, Andrew McCutchen."

"I like infield better."

"Everyone does. You'll probably play second, too."

That day, we had our best pitcher going, and the umpire had a decent strike zone, which played to our advantage. We scored six in the top of the first, and it looked like it might be a blow out, but the Tigers battled back. The bottom of our order went down quietly in the second, and the Tigers plated two more, but our center fielder caught a fly ball to prevent a big inning. I took the catch as a good omen.

We were up 12–4 by the bottom of the fourth. Then the Tigers started taking pitches and drew a few walks. They had one really good hitter, and before she came up to bat, we visited the mound. We told our pitcher to throw everything as high as she could. If they were going to take pitches, they could take four more.

The Tigers coach, who seemed overly friendly before the game while he was lining and watering the field, yelled over to our bench, and their fans started getting into it. We got two outs and were about to get out of the inning when they hit a ball back to the pitcher with runners on second and third. Instead of throwing to first for the last out, the pitcher threw home, and our catcher caught the ball but didn't tag the runner. Then the catcher overthrew third, the left fielder overthrew second, and everyone scored.

We were up four after four.

We went up six on a two-run home run to dead center that rolled forever in the bald outfield, but in the bottom of the fifth, we couldn't get an out, and the bottom of their lineup kept taking pitches to load the bases for the top of the lineup.

It went back and forth with great plays. After six innings, we were up 16–15.

The max runs per inning rule was removed in the last inning, and we took advantage of it. We batted around all twelve girls and scored eight runs. Gracie had two hits. The Tigers scored three in the bottom of the inning and had runners on, but our pitcher worked out of it.

Final score: 24–18.

The trip home from Hazelwood was quick but celebratory. I was relieved we hadn't lost, and Gracie was already looking forward to her game in Lawrenceville the next night.

We played Brookline in the playoffs and lost by fifteen runs. Gracie got a hit in her last at bat for Shadyside, and we generally looked good at the plate as a team. But Brookline hit balls off the fence all game, made all the routine plays, and we were outclassed. The Shadyside softball season ended with two hard-fought wins and an early playoff exit.

July 2019
SALTING FALCON AND A FRIEND ON TOPS

My friend Bruce was looking for a job, and the new team Will was forming at HCL was looking for new people to help improve the algorithm that selected images for product pages, filtering out inaccurate, offensive, and mismatched content. The team was primarily new hires but included a few internal transfers as well. Bruce wanted something more stable than adjuncting for the summer, so I sent him a link to apply. He was hired and immediately joined the organizing committee. I was very pleased with how it had worked out. I was able to help both a friend and the organizing effort. And Bruce and I were finally able to get lunch together.

Falcon was secluded into two offices near our pod. Their work was even more boring than ours, and the company tracked their productivity in half-hour increments. There was no benefit to doing a good job. You couldn't leave early if you finished enough work, and you wouldn't get paid more if you exceeded your required pace.

A new hire, Riley, was brought on to work on the taxonomy team (TOPS). Riley joined the committee the same week as Bruce. There were thirteen of us now for a potential bargaining unit of eighty-five. We worked fast. By mid-July, we were on the cusp of handing out cards. Our committee meetings were upbeat, and the days at work had a nervous energy about them.

July 2019
CARDS

Before every meeting, I thought we'd burn out or that someone would rat on us and we'd get fired. But we kept going. We hit benchmark after benchmark, and soon it was time to go public. By the end of July, we exceeded our threshold of pro-union workers, and we planned a party at a bar off Hamilton. We would have union cards there. We made flyers to hand out at the office; they contained a summary of our labor rights on one side and the party details on the other.

The night before the event, our meeting was energized. We were so close, and it had all come pretty easily so far. All this time assessing and inoculating—preparing the staff for union-busting talking points—and management hadn't said anything yet.

"What, is HR gonna come in and tell us how well they treat us?" Will asked at the meeting. "No one is going to fall for that."

Management seemed mostly oblivious to our plans. If they had heard anything, perhaps they figured unionization was too wild of an idea to take seriously, even though some of them made less than the workers they supervised. Laura was leaving soon. Donald was slated to take over duties as team lead and lead analyst again. He was also running a team in Kraków. I hadn't seen Nancy in weeks. We didn't know what to expect.

The next morning, I got into work, and Will nodded at me, pointing toward the cabinet behind his desk. I walked over, grabbed a few flyers from a folder sitting on a nearby shelf, and stuffed them into my bag. On the other side of the office, Riley had a stack to hand out as well.

We'd had a miscommunication about our strategy. It was assumed management would find out because that's usually what happens when cards circulate, but it was up to us to decide how careful we were going to be in the office. Most of us remained discreet, but a few committee members started handing out flyers openly in the common areas.

I met Will out on the patio overlooking Target. We sat at a plastic picnic table with our laptops.

"Maybe they won't do anything. Maybe they'll voluntarily recognize our union," he said.

"It's hard to say."

"I'm stressed."

"When I get this anxious, I get paralyzed," I told him. "I don't know what to do, I'm so worried."

"I'm the opposite," Will said. "I try to fix everything."

We talked shit and hypothesized about how the rest of the campaign might go. Then one of us had to leave for a meeting.

I did enough work that day to keep my weekly numbers from dipping too much. I left the office, picked up Gracie from the Boys and Girls Club, and waited for the babysitter before I went to the card-drop party.

As I drove down Hamilton, I saw Kate and Alice walking away from the event, back to their cars. I assumed they'd gone down to sign cards, and I was relieved. As of that morning, neither of them had committed one way or the other yet. Over the last few months, I'd gotten to know them better, and I liked them both. The two of them had been tight-lipped during assessment. They were paranoid about management finding out and worried about retaliation. When I drove past, I waved, but they didn't see me.

I parked and walked to the bar. It was a newer place with a lot of outdoor seating. The music was good, and it smelled like barbecue and beer. There was a huge crowd from work around one of the big wooden tables. People I only knew in the context of the dim, open office plan were there, flush with laughter and alcohol.

I ordered a soda water and sat in a booth, where I met Damon and a bunch of other USW staffers. I signed my union card and put it in the box with the others. It didn't feel real. I looked around and took everything in. I wanted to remember it in case things went bad. It felt like we had a chance to revitalize the labor movement, to start a union that would promote social justice and the fight for the most vulnerable people in our community.

"You did this," Damon told me.

"I helped get the ball rolling," I said.

"It's OK to take some credit."

"It feels great," I said, but I liked that I had been given the chance to deflect the credit. Realistically, I got more than I deserved. I was in the right place at the right time. I didn't consider myself a natural workplace leader, but I'd settled into my role as agitator and advocate. I wasn't afraid of management or getting fired. Mostly, I was happy we were all there and feeling optimistic. Everyone wanted a bigger say in how things were done.

We sat and talked about work. The committee welcomed everyone in, and a few more people volunteered to help. We met each other's significant others. We commiserated about our jobs, bad bosses, and HCL's ineptitude. Everyone had stories. We had grievances large and small, and we talked about how we felt we'd been manipulated by our bosses and by Google.

When we counted the cards at the end of the night, we already had over 30 percent. We had met the minimum, legal threshold to file for a vote, and we were halfway to our goal. USW staff said, proportionately, it was the biggest turnout for a card drop event they'd seen in years.

August 2019
CARD COLLECTION

Back at the office, we passed union cards in novels and notebooks and cigarette packs. We hid in stairwells and parking garages. We even met people beyond the backstop of Gracie's softball field near the concrete paths that ran between the condos behind the office. We passed cards in the break room, the bathroom, and the lobby. Organizers had discussions in German, Spanish, and French within earshot of bosses who had no idea what they were saying.

We had a good system. Each of us chose individual coworkers to talk to and made time during the day for one-on-one conversations. After getting cards from all the definite yesses first, we worked our way through each of the teams. We continued assessing and inoculating. By the first week of August, 50 percent of the workforce had signed. Management still hadn't taken any action.

Will spent hours on the phone with coworkers who were on the fence. He got mixed but positive results. I stayed with the people I knew best and was becoming friends with. If workers were supportive of the effort already, I tried to get them more involved. I had to have some difficult discussions, too, with people who were virulently anti-union. I did my best.

As we neared our goal, every signed card carried more weight, and every conversation that went poorly felt like the end. The emotional swings were draining. We were either going down in flames or changing the face of the labor movement. There was no middle ground. When the first vocally anti-union worker emailed Jeffrey about what we were doing, Jeffrey didn't respond. The lone holdout on Falcon got scared and said something to Nancy, but nothing happened.

Nancy still worked from home most days, and Jeffrey stayed in Minnesota. Our HR rep worked from North Carolina and was never on-site either. Even after the flyers and the event at the bar, nothing happened. Even as the union formed around management, even when they were told directly about it, they did nothing. I'm sure some of the supervisors on-site had some idea of what was happening. Maybe they knew what was happening and didn't care. Maybe they'd heard the whispers but didn't think we would follow through with it. The conditions were ideal. We just had to stay careful.

By the last week of August, over 60 percent of the workforce had signed cards—nearly enough to file. But then the goalposts shifted. USW's lawyers reviewed the employee list and asked Damon, "Why are there fifteen people with the title *lead analyst* for a bargaining unit of eighty-five? That's twenty-five supervisors, including the team leads."

Some unions won't file for an election unless they get 80 percent of the workforce to sign cards. Our threshold was 66 percent, more than twice the legal minimum. The conventional wisdom says the reason that number has to be so high is that organizers anticipate losing 10 percent during the anti-union campaign, and it's not worth the resources to file for a vote if there isn't already a good chance of winning. This isn't the strategy in every successful organizing drive, but for our workplace of ninety, it made sense. The size of the majority also comes in to play when workers need to take direct action against the company. Direct action won't work without a strong showing of solidarity. If we all decided to wear red shirts one day to support the union and ten people wore them, we'd look weak—they might not even notice. If there wasn't a big pro-union margin, it would also be easier for management to get a decertification campaign started down the line during bargaining. That's always management's plan after a union win in an election—stall for a year and then try to decertify. The more support we got now, the more leverage we would have in a number of situations. Still, we'd been stressed out for months, working hard, spending time away from our families and other interests, and we wanted to file. We were so close, and now we'd have to get six more signatures instead of two, and the last few cards are always the toughest. It might not seem like much, but it felt deflating.

Time was becoming an issue. If management got to a point where they could no longer ignore what was going on and finally had to take action, they could do any number of things: illegally fire organizers, interrogate people, hold one-on-one or group meetings in which they'd try to scare us out of organizing. Also, a dozen pro-union workers were set to leave in September, and we wanted to file while they were still on the payroll. We had until the Friday before Labor Day, or we'd have to start all over again.

We all thought we would win if we just got it to a vote. I fielded multiple calls from committee members who asked if we could file short of our goal. I asked Damon if we would file without a supermajority.

"Maybe if we're one card short, but not six," he said.

—————

At our next meeting, we made plans to talk to everyone who had not been assessed as a four. Will got cards from a couple of the lead analysts. He cut deals with some of them about keeping their identities a secret from the rest of the committee, even though that was basically impossible. We never told anyone outside the committee who had signed and who hadn't unless a signee specifically said it was OK to do so. We didn't use anyone's email who hadn't given it to us, and we didn't text anyone who hadn't given us their number. It would have been bad if it got around that we weren't protecting people's privacy. We were taking a risk being openly pro-union, and it wasn't our place to impose those stakes on anyone else. Plus, when it came time for the bosses to counter-organize, the less they knew, the better. And the less everyone outside the committee knew, the less management would know. We needed to use our numbers to our advantage to show people on the fence we had support, but some folks who signed early were getting antsy.

As the pressure increased, a few of us worried Will was going to do something rash that would sink our chances. Will could be divisive, but he secured a ton of signatures. Every meeting, when a new person showed up, he retold the same jokes about how HCL obfuscated, escalated, and ignored all our requests. He was vocal about his distrust of management and repeatedly told everyone how much he'd been fucked over. It could be grating, but he could also be really funny and engaging when he eviscerated the anti-union messaging and HCL management. Sometimes it was good to have a friend who would tell you what you wanted to hear and indulge your worst impulses. He'd text me late at night while we were both awake and feeling anxious about the numbers:

Will: Hey Ben, I have a bad idea.
Me: What?
Will: How about we start a rumor that Ernst is a Nazi?
Me: No.

Will: Do you think Donald's briefcase is full of dismembered doll parts?

Me: Absolutely.

But he was perpetually on the verge of a meltdown, questioning the whole ordeal and trying to do everything himself. When we tried to impart the importance of delegating tasks so everyone felt like we had ownership over the process, he'd be OK for a while, but then he'd backslide and start trying to run meetings again, telling us stories about our current middle managers from years ago that didn't have much relevance and weren't that entertaining. A few other committee members had a really hard time with it: to them it felt like another instance of a white guy trying to control everything and not giving anyone else space to talk. Sam stopped coming to meetings as often if they knew Will would be there. Paige and Sarah were also getting fed up. Carl and I tried to give him another way to vent outside our committee meetings. We almost had to have an intervention.

The last week before filing was hard on everyone. I was so sleep deprived, I practically folded in on myself. Among the last of the threes were my teammates: Kate, Carolyn, and Alice. George eventually got Carolyn's card after she had ignored twenty-seven texts from Will.

August 2019
ALICE AND KATE ON THE BALCONY

I was sitting out on the balcony one afternoon when Will, Alice, and Kate walked out and sat at the table at the far end of the patio. Kate and Alice still hadn't signed a card. They were too worried about Nancy finding out, and they were put off by our brazen flyer distribution. They were close with George, though, and I'd been becoming better friends with them in recent months. I felt comfortable enough talking to either of them when we got down to the end, but I figured George would talk to them first and would do so at the right time. They respected him.

Will went into his speech about why we should have a union and why we needed Alice and Kate to sign cards immediately.

"I can't sign it," said Kate. "I'll support it if we win the vote, but I can't. Nancy could find out. She'd fire me."

"It's completely confidential," Will said.

"We tried this years ago with the cafeteria workers, and it didn't work," she said. "We don't know what's going to happen with those cards."

There was no changing Kate's mind. She had good politics, and I liked her and Alice. I was sad they hadn't already signed.

Will kept at it.

"Why don't you want to change things?" he asked.

"I'm not comfortable talking about it here," Kate said.

I looked at Alice and shook my head. I moved around the corner to keep watch. I stood and stretched and looked at the door. There were a couple pro-union people sitting in the library, surrounded by ornamental books that had been chosen for the colors of their spines. I stared through the glass door and focused on one of the books while the uncomfortable conversation dragged on behind me.

Will started sobbing.

"But we need to get to two-thirds or we won't even get a vote. It's so close."

"I can't risk someone finding out that I signed and it getting back to Nancy."

"Why? We need this. We've been shit on for so long. You know what it's like. It's against the law for her to fire you for signing a union card."

Alice and I looked at each other in stunned silence while Will and Kate argued.

"Please, I'll meet you after work," Will said. "I won't tell anyone."

"I can't. I will support it if you win the election, but I'm not ready to take that risk."

"This is our only chance. Alice, you had cancer and they fucked you over. How can you not see this?"

"I don't know, Will," she said.

I left before the conversation ended. I didn't know how Kate and Alice would respond, but we needed them on board, and this wasn't the way. I felt like I should've tried to diffuse the situation, but I didn't want to somehow make it worse.

Later that week, I ran into Alice and Kate in the cafeteria.

"I had nothing to do with that out on the porch," I said, "and I'm sorry."

"It's fine," Kate said. "It's just Will. He's like that."

"Not your fault," Alice said. "It's really OK. Will is emotional. I get it."

Apparently, issues like this were common in organizing.

There are people who try to do everything and others who want things done a different way but who won't come to meetings or do any work. Some people think their way is the best way and are inflexible about it. We had to keep everyone on the committee engaged and sharing the responsibilities.

———————

That week, I had a reading at White Whale Bookstore in Bloomfield. I don't remember what I read, but it was a great night out. I realized how much I missed my friends in the literary community. I had to remind myself that I was organizing so I'd have a job with more flexibility, so that I'd be able to write and take care of my daughter. I often lost sight of that—I'd get lost in the spreadsheet, looking at updated notes, thinking of ways to run into people so I could have an organic conversation about the union, or worrying about Will doing something over the top. I tried to get to a place where I would be OK if we didn't get enough cards, but I found it difficult. If we failed, I'd start applying to jobs again. I'd eventually find something, even if it meant bussing tables for a few months in the meantime.

August 2019
ANOTHER MEETING ON THE BALCONY

The next week, George sent Sam and me a message, asking us to meet him out on the balcony overlooking Penn. Sam and I sat and waited for him at the plastic picnic table.

"I'm worried about Will," I said.

"So am I," Sam said.

"Do you know what George wants?" I asked.

"I have no idea."

We were commiserating about the expanded bargaining unit when George came out through the library and sat with us.

"I'm leaving," he said. He'd put in his notice. He was going to teach high school in New York.

Sam and I were crushed.

"Regardless of how it turns out," George said, "we've done something great." We all held hands. "If we don't get the cards, then this isn't what people want. If we lose the vote, it wasn't meant to be."

I could barely speak. I couldn't look at Sam because if they cried, I would cry too. I stared at a scratch on the table. George and Sam left, and I sat there deflated, listening to the traffic on Penn. Then I went for a walk around the baseball field and tried to collect myself.

———

The day after George told us he was leaving, Nancy put in her notice. She had accepted a job as a manager at an Amazon warehouse in the suburbs. Her last day would be the Tuesday after Labor Day. The more conspiratorial of us wondered if she knew in advance that our jobs would continue to go to Kraków and Gurgaon, India. Maybe we were closing in on perfecting the algorithm that would fully automate our jobs. Maybe Nancy just wanted to force workers to piss in bottles during their shifts.

Our work had gotten worse in the short time I'd been there. The standards dropped, and we had to think less and less. On Catalogs, our primary work devolved from editing and analysis to mindless mouse clicking. Many of my coworkers on other teams did the same work as full-time Google employees for a fraction of the cost. Some teams did extremely difficult technical work, some matched pictures, and some did both. Some weeks I did nothing but look at product pages and decide if they were up to standards or not. The work was engaging and challenging just often enough for it to be frustrating when it wasn't. The job was killing my brain. I felt like I was getting dumber.

August 2019
HOUSE CALLS

Our supporters in the office were asking why we hadn't filed for an election yet, and we had to answer a bunch of procedural questions about the size of the bargaining unit and about why we had to wait. We tried to project confidence while we worried about getting the last of the signatures. We were two cards short with two days to go. It was time for house calls. We would either get the last cards, or we wouldn't file.

I was less apprehensive about house calls than I was about approaching someone to get coffee at work. All the pretense was gone with a house call. Plus, we went in pairs, so it felt less intimidating.

I was paired up with Sam; they agreed to do most of the talking, and I would interject if necessary. We had an address for one of the Annas (there were several in the office). I didn't know any of them, but Sam did. The Anna we were visiting had been promoted without getting a raise, and we weren't sure if she was a supervisor or not. We hopped in Sam's electric car and buzzed off down Route 22. On the way, they told me organizing stories that ended with them getting chased off porches.

"Do you think Damon will file if we're like two cards short?" Sam asked.

"Maybe," I said.

"I don't get the thing with the leads. How do we have twenty middle managers?"

"How's Falcon?" I asked.

"Terrible, but I don't have to deal with Donald."

A year earlier, Sam and I had met on the couch outside the game room, excited for our new jobs "at Google." Now we were driving around aimlessly, lost in East Pittsburgh because Google Maps hadn't updated and we couldn't find the side road that led to the apartment we were supposed to visit to try to get a union card signed by one of our most buttoned-up coworkers, who was probably a manager anyway.

Eventually, we found the right street. It wound up the side of a big hill and spit us out in front of a sprawling apartment complex that looked like a hotel.

"It's an old folks' home," Sam said.

"Maybe she's taking care of a relative."

The automatic doors opened, and inside, there were muted TVs and several senior citizens sitting in big plush chairs. Someone drove by on a scooter. Nurses walked past another big TV on the wall that was playing an infomercial for the residence itself. The hearth was full of fake logs. It didn't look promising.

We stood by the empty receptionist's desk for a minute and then wandered around the lobby. I poked my head into an empty room where a caterer was breaking down a banquet table, and he directed us to the night manager's office down another hallway. The apartment number we were looking for was in the 400s, and there were only three floors.

We sat in another waiting room before the manager came in and told us we were in the wrong place.

"Thanks for your time," Sam said.

We drove back to Bakery Square. We went home empty-handed but with a story to tell. Hopefully, the other house calls would be more successful.

August 2019
IN MEMORY

It had been a full year since Jane had died, and I was no closer to processing my grief or coming to terms with the conflicting feelings I had about our relationship and her death. Her funeral service had taken place late in the summer. I put on my wedding/funeral shoes, a white button-down shirt, and an eight-year-old pair of black slacks I still had from when I bussed tables, and I drove up to the park in Zelienople. Gracie was already at her grandmother's for the weekend, and I met them at the service. Jane's mother didn't want anything too outlandish, so there was a picnic and a brief service. Jane's uncles on her mother's side didn't attend. Her dad had held his own service at his church in Emporium.

Jane's family relationships had been strained by addiction. Her boyfriend and his parents had only known her for the last couple years, and it seemed like there was a bigger contingent from his side of the family than there was of Jane's. Since I'd known her, she'd been kicked out of her mom's house three times and her grandmother's twice.

Jane's brother and sister were there, talking to people I didn't know. Gracie kept close to her grandmother and her aunt. I recognized the social worker who'd been assigned to Jane's case. She was shocked. We were all shocked, more so I think at the timing than the outcome.

"I don't know how I didn't see it coming," she said. "I feel terrible. You have such a good kid."

"She fooled me for years," I said. "I thought she got it this time, though."

The social worker once recorded in her report that Gracie had not had her jacket when I took her to meet Jane in the park.

Another time, she noted that Gracie's hair wasn't presentable when I dropped her off. I did my best, but it was hard.

Whenever I brushed Gracie's hair, I thought about an essay by Michael Paterniti in *Esquire* about Thurman Munson, the New York Yankees catcher who died in a plane crash. The essay included an image of Munson slowly brushing the knots out of his daughter's hair. Paterniti returns to that tender image throughout the essay about the toughest catcher of his era. Munson had been the captain of the Yankees. When he was growing up, his father was insanely hard on him when he was around, but he spent most of his time driving a truck. Growing up in New Jersey, I thought Yankees fans exaggerated how great Munson was, but if anything, that article made me think they undersold him. It's the most beautiful piece of sports writing I've ever read.

Sitting at the pavilion on the day of Jane's funeral, I thought about that essay, wishing I could write or play baseball half as majestically as Paterniti or Munson. Mostly, I felt grateful I was alive and could be there for my daughter. It was somber, but not as sad as you might expect when a thirty-three-year-old dies. Earlier that summer, the brother of a childhood friend had died, and I had gone to his funeral. He was in his forties when he'd had a heart attack. My cousin died in a hotel pool in Texas a week later; he was only twenty-seven. Mourning the loss of someone so young is hard. Everyone gathers and grieves and tries to get through it. Maybe it's easier when it's clearly drug- and alcohol-related—or maybe it's just less of a surprise. I don't know. But what struck me about Jane's service was that none of her old friends from NA were there. They'd either died, were in jail, or had gotten clean, and Jane burned them for money so many times that they didn't feel the need to come to the funeral.

When I first met Jane, she had a bunch of sober friends. Only one of them was still clean now, and they didn't go to meetings anymore. I talked to them after Jane's death, and we reminisced about those days of early sobriety when we were happy just to be alive, when that was enough. We were both glad the other was still sober and doing well. But there were so many people I met through Jane that I saw for a few months, or less, who disappeared from the rooms or who dipped in and out, over and over, unable to get enough distance between themselves and drugs and alcohol to stay quit for good.

There was a microphone at the edge of the pavement by the front benches and a PA set up for the pastor—or whoever the spiritual guy was who was going to speak. I walked out to the playground and smoked a cigarette with Jane's siblings. They told me Jane had relapsed years earlier at her mother's house. She had stolen a bunch of her brother's pain meds and overdosed in his bathroom while Gracie was asleep upstairs. I hadn't known. All those hopeful nights I had let Gracie stay up there even though I was pretty sure Jane was fucked up—but couldn't prove it—backfilled into near catastrophes of needles and shady houseguests. Now, after almost ten years of worrying, Jane was dead, and I couldn't figure out how I was supposed to mourn her.

Someone I didn't recognize tapped the mic. It was time to start.

"Mr. Parsons will say a few words, and then anyone who wants to speak can come up and talk about Jane."

I sat with Gracie and her grandma and tried to organize my thoughts.

August 2019
CARD DEADLINE

The last Friday in August, Sam talked to Kate outside by the gated community and struck out again. I checked the group chat frantically. I thought we were done. We'd made plans to meet for pizza in Bloomfield to either celebrate how far we'd come before filing the cards or to commiserate about our failed but valiant effort. Then we got a message from Paige. She had driven out to Alice's place and gotten her to sign a card. We were still one short. I stayed late to collect any extra cards and bring them to dinner. When Donald left, I went upstairs and played pinball for an hour. I was about to leave for the night when Sam messaged the group chat.

"I got it!" they said. "Kelly signed a card."

Kelly had been a four but wasn't vocally anti-union. We hadn't contacted her in a while, but Sam had asked her randomly because they were both staying late that day—and it worked. We had our supermajority. The party that night was celebratory but sad. Delia was leaving. George would be gone sometime in September. Several other supporters were leaving for grad school or new jobs. We stuffed the jukebox full of money and I sat with a soda water while everyone else drank and ate. We began to look ahead at what might be next.

Damon filed the cards at the regional NLRB office the next morning. After the long weekend, the company responded to our petition. They had hired Ogletree Deakins, who was known as one of the biggest union-busting law firms in the country. The election would take place on September 24.

September 2019
MANAGEMENT RESPONDS

Toward the end of the nineteenth century, the Second Industrial Revolution led to the Gilded Age, which until recently, was the era in American history characterized by the greatest level of wealth disparity. New technology led to massive increases in productivity and more dangerous conditions in factories, mines, and mills. Railroads and automobiles were new and exciting. Capital was able to produce at unprecedented rates. Worker-led labor organizations emerged in greater numbers as a result of a new wave of strikes. During this era, workers demanded an eight-hour workday, the end of child labor, and other basic rights we take for granted today. In many cases, like the Haymarket Riot and the Pullman Strikes in Chicago, when workers asked for better conditions, they were met with violence.

Early on, there were two basic types of labor unions: craft unions and industrial unions. Craft unions were made up of skilled workers and limited to one sector like carpentry. A number of them joined together to form the American Federation of Labor (AFL) in 1886. Industrial unions spanned entire plants and various industries. They included skilled and unskilled workers (I don't believe in the idea of dividing labor into skilled and unskilled categories, but those are the common terms), and they eventually made up the Congress of Industrial Organizations (CIO).

The CIO formed in 1935, when a group of unions left the AFL because it didn't want to organize unskilled workers along industry lines. The CIO was more radical; its leadership included members of the Communist Party USA. Communists historically have been the most effective labor organizers, and they have also opposed segregation and racism.

The AFL and CIO eventually merged again to form the AFL-CIO in 1955. Fifty years later, another coalition of unions formed when the Teamsters and the Service Employees International Union left the AFL-CIO and started Change to Win. The split was caused by ideological differences about decision-making and strategy.

In the wake of the Great Depression, the National Labor Relations Act granted American workers the right to formal unionization, collective bargaining, and other long overdue protections. The NLRA does not include provisions for independent contractors or for agricultural or domestic workers, however. These omissions disproportionally affect people of color and have not been addressed in a meaningful way on a national level since they first came into existence almost a century ago.

As soon as the NLRA was passed, corporate interests started attacking it. The Taft-Hartley Act, passed in 1945 at the height of the postwar labor boom, gave back to management a great deal of the power they had lost with the NLRA. Among other things, Taft-Hartley allowed companies the opportunity to make their case against unions during union drives. It also made it illegal for Communists to hold union leadership positions, and it gave states the right to enact "right to work" laws. Labor rights have eroded as a result.

During World War II and through Vietnam, the labor movement generally made bad decisions. Unions sided with prowar Democrats and perpetuated Red Scare policies that hurt organizing and saw the expulsion of many radicals from their ranks. And though labor's ties to organized crime are exaggerated, those also existed.

In 1980, President Reagan fired striking federal air traffic controllers, escalating the fight against unions. It was around this time that anti-union law firms and consultants grew in popularity. A cottage industry of these law firms and consultancies sprung up and eventually grew into a big business, all built around keeping unions out of workplaces. They helped companies stay within the letter of the law established by the NLRA without really honoring their employees' rights to form a union without managerial interference.

In the twenty-first century, new technological advancements have allowed productivity to soar. Income inequality is now greater than it was during the Gilded Age. My generation has faced harsh economic downturns in both 2008 and 2020. Wages have not kept up with productivity or inflation and have been mostly stagnant since the seventies. The last time things were this bad for so many working people, Franklin Roosevelt passed the NLRA to stave off socialism, and the labor movement boomed.

We started our union with a lot working against us. The NLRA is so watered down by now that it isn't even punitive. If management illegally retaliates during a union drive, all they have to do is reinstate illegally fired workers with back pay, post a notice admitting they broke the law, and promise not to break the law again. The National Labor Relations Board (NLRB), the governing body in charge of enforcing the NLRA, is understaffed and overworked. Republican presidents gut it when they're in power. Democrats undo some of the damage but not enough to truly strengthen the institution. Obama did not pass robust labor reform when he had the chance. The Protecting the Right to Organize Act (which includes heavy fines for corporations and, potentially, individual supervisors who violate labor law, along with new items on the list of illegal union-busting actions) is currently at the mercy of the Senate parliamentarian as Biden's cabinet tries to pass some of its key provisions through reconciliation.

Under current law, management doesn't even have to come to an agreement on a first contract during bargaining. Corporate law firms advise management on how to stall and anger workers in order to turn them against the union so that they can then astroturf a decertification campaign and vote the union out. The only way to get a first contract is to leverage a credible strike threat, unfair labor practice (ULP) charges, or bad press—and preferably all three—against the company.

At HCL, we did not have a credible strike threat. Our work had been outsourced to Kraków for years; that was one of the things we hoped our union could help curtail. We had the press, but ULPs, which violate national labor law, were our most important point of leverage. Typically, they come in the form of retaliation in response to unionizing or other collective action by workers. ULPs, sometimes referred to as board charges, range from changing the amount or type of work employees perform to illegally firing someone. While HCL wouldn't be penalized for ULPs, they would theoretically have to reverse their illegal actions if they were found to be in violation of the NLRA. A litany of charges would make them look bad, which would then make Google look bad at a time when they were trying to stay out of the news for anything related to labor, especially as it pertained to TVCs. An NLRB hearing would also cost HCL hundreds of thousands of dollars on top of the hundreds of thousands they were already spending on lawyers.

As soon as we filed for an election, USW lawyers trained us to spot actions that violated labor law and could lead to board charges. The other tool we had was direct action. We couldn't strike, but we could visibly support our coworkers and our union in other ways, mostly through wearing stickers and pins. Unions love stickers and pins. We could also hold rallies, organize walkouts and work slowdowns, and circulate petitions. We could still make demands of management (which we did successfully when the pandemic started). We would have to keep organizing during negotiations to show solidarity and keep morale up while management tried to stall. But before we even got to collective bargaining and negotiating for our first contract, we had to win our election.

After we filed our cards with the NLRB, management did their best to bust our union. In addition to hiring Ogletree Deakins, they also brought in a celebrity union-busting consultant, Eric Vanetti. Ogletree Deakins is a typical management-side employment law firm. They provide a number of services that help corporations exploit workers. They defend companies accused of crimes such as OSHA violations (including COVID-19 protections), denying worker's comp, violating environmental regulations, sexual harassment, wage theft, discrimination, and illegally denying unemployment claims. Firms like this are on the side of some of the most powerful corporations on the planet. They work for people who would still be rich if they paid out every lawsuit brought against them, even the few frivolous ones.

Vanetti worked for a consulting firm called the Labor Relations Institute. He had been in the documentary *American Factory*, which happened to be in rotation on Netflix during our union drive. In the documentary, he was brought in during a union drive at a glass plant in Akron, Ohio. Often, he gave talks and presentations to workers who worked harder and made less than we did. People like him are typically paid around three thousand dollars a day plus expenses.

During union drives, management often brings these consultants in to hold captive audience meetings. The meetings are supposed to be voluntary, but they really aren't. They're held on the clock, so workers get paid to listen to anti-union talking points. This is framed as the company getting a chance to present facts and tell their side of the story. These meetings don't necessarily turn people against the union on their own merit, but they serve as a framework for management to do their own assessment of the workforce, an opportunity to try to convince workers who are on the fence to vote against unionization. They do this by forming their own anti-union committee of workers who are already against organizing. That was the hardest part of the whole thing for me. A small but vocal group of our coworkers sided with the boss and eventually went after us and our union.

Legislation like the PRO Act would make captive audience meetings illegal. Without captive audience meetings, the narrative of unions as confrontational would lose credibility. This narrative is largely created by the consultants and the lawyers who tell management not to answer work-related questions or perform their basic duties for fear of the union. In the vast majority of cases, they tell management not to voluntarily recognize unions. Unions are othered rhetorically by the consultants and lawyers. That's a standard talking point, "The Union is a third party." The Steelworkers were described similarly, and we were urged to trust the company that had never cared at all before we wanted to unionize.

We didn't want a confrontation. We wanted more time off. We wanted equal pay for equal work and real raises. We wanted to work from home and to have dedicated sick days. We wanted to exert our legal right to bargain over these things collectively.

The minority of workers who were vocally anti-union accused us of gambling with their future and "risking everything" for what would likely be minimal gains. The anti-union group included both the second-highest paid and the second-lowest paid workers at the company. Most of the anti-union employees made above the median salary. Some just didn't want to rock the boat. Sure, the company had issues, the thinking went, but most companies did, and it could always be worse.

Usually, when a company hires a firm like the Labor Relations Institute, the union loses. After maintaining our support, getting cards signed by two-thirds of the workforce, and filing them at the last minute, we would have to fight a multinational corporation with unlimited resources. We were well prepared, but I don't think we quite understood what was about to happen.

September 2019
UNION BUSTERS AND RETALIATION

The Tuesday after Labor Day weekend, Jeffrey taped the NLRB notice to the whiteboard near our cluster of desks—the one that usually held the question of the day. I watched as workers gathered around it. The notice said that we had filed for a union election under the National Labor Relations Act and that the election would take place in three weeks at the library in East Liberty. If we won, USW would become our exclusive bargaining agent, and we would begin negotiating a collective bargaining agreement with HCL.

A few anti-union folks near our pod voiced their disapproval when they saw the notice. They complained loud enough for us to hear it. I looked across the office to Will, who was smiling. George was chatting with Kate, Alice, and Carolyn. One of the more vocal union supporters who sat behind me was openly cheering. Jeffrey had to come all the way from Minnesota to post that notice.

Before we could say anything, everyone was called into separate meetings with our team leads. I sat with the rest of Catalogs in a conference room. We waited for Donald and Nancy for what felt like a long time, though it was probably only a few minutes. While we were sitting there without our supervisors, wondering out loud what was about to happen, we asked everyone to bring questions to someone on the organizing committee and not to our supervisors.

When they entered the room, Donald sat in the corner, away from everyone. Nancy went into an anti-union tirade, hitting all the typical talking points.

"You don't need a union, in my opinion," she said. "I believe in unions, but not in this building. My best friend is in a teachers' union. She needs it, but we don't. We have everything we need here. Free food, laundry, snacks, and video games. If you bring in a union, we don't know if we will be able to get our contract with Google renewed. You might have to go work out by the airport. Unions won't let us be as flexible as we need to be to do our work for Google. Besides, by 2022, we will get more sick days. We offer tuition reimbursement and health insurance. Do you want to lose your health insurance?"

Nancy had seemed sympathetic when I had suddenly become a single father. She lamented how unfortunate it was that she couldn't do anything to help me but that her hands were tied. Now she was furious that my coworkers and I wanted to actually improve our situation and the situations of our families. It was Nancy's second to last day at HCL before she left for Amazon. We were told later that she got a bonus for giving us this speech on her way out the door.

Donald stepped in.

"If you have any questions, please come to me or Nancy," he said. "We also have a visitor who will give you some unbiased information about unions. You can take as much time as you want, on the clock, to talk with him. The first meetings will be held later this week."

We were dismissed. Everyone was pissed. Luckily, we'd told everyone what management was going to say, so no one was surprised. Carolyn, Alice, and Kate—who were all ambiguous about their support before Nancy's tirade—were now angry and vocally pro-union. We all took a break out by the pond.

"That was bullshit," Carolyn said. "She's leaving, why does she care?"

"What a joke," I said.

"Amazon is perfect for her."

"They're bringing in a consultant."

"This is the fastest they've ever responded to anything."

Any thought that HCL was too incompetent or cheap to fight our union was gone. We had caught them flat-footed, but they would gather themselves and fight.

September 2019
THREE WEEKS UNTIL THE ELECTION

Once Nancy left, her position was temporarily filled by two supervisors, Ernst and Larry, a classic good cop, bad cop duo. Larry played the good cop. He smiled a lot, stood about 5'5". He looked like a cannonball. I'd see him outside vaping on occasion. Ernst, the other team lead, was from somewhere in Europe and had a goatee and a background in psychology.

A few days after the first meeting with Donald and Nancy, Eric Vanetti arrived. Thankfully, Google wouldn't let him in the office. I was so paranoid, I once thought I saw him standing by one of the coffee makers, but it was a different older white guy in a sweater. Vanetti held meetings in the conference room at the Spring Hill Suites, which was in the strip mall right across the parking lot. Everyone was sent calendar invites to go talk to him. We made sure we sent pro-union people to every meeting to counter his talking points.

HCL had had a chance to preempt a unionization effort by providing fair pay and equitable working conditions. They could have fixed things when they got the Google Shopping contract and took over all the TVCs. They chose not to. We were called divisive for trying to make things better.

Management started the fight before we even organized. Right-to-work laws, our shitty health care system, stagnant wages: all of them are acts of violence. Organizing was an act of self-preservation, not a sucker punch. Since its founding, the United States has promoted individualism at the expense of the group. If we failed to get rich, or at least join the middle class, it was supposedly our fault. Collectivism is portrayed as weakness. Anything to the left of Nixon, politically, is called Communism and denounced as a threat to freedom. Capitalism keeps just enough of us just content enough that disrupting the status quo seems impossible and not worth the trouble. Most people don't have the time or energy to rally around any number of causes that need support. Our country turns its citizens into islands, while a handful of people have become obscenely wealthy.

We saw some of these broader cultural arguments against collectivism play out in real ways in our workplace for three weeks as management brazenly tried to bust our union. It was a microcosm of the culture wars that inform most of the anti-leftist sentiment.

At our office, there were typical right-wing folks who were vocally and virulently anti-union. There was at least one guy who, during his time at Google, went from normal tech worker to white nationalist conspiracy theorist who sent articles from places even crazier than *Infowars* to his coworkers. There were the "I'm not political" people. But there were also people who were proud, anti-Trump, "Third Way" Democrats who thought we should be nicer or find another way to improve conditions without taking such a big risk.

I walked over to the info session with Kate, Alice, and Carolyn. As we made our way across the parking lot, I had butterflies in my stomach, like I was getting ready for a baseball game. It smelled like cement and car exhaust. I was sweating—worried I'd say too much or too little, or that I'd say the wrong thing.

Vanetti had gray hair and dressed like Mr. Rogers. Every session started the same way, with him sitting on a desk in his cardigan, ripping open a LUNA Bar.

"Sorry," he said. "I worked through lunch."

Very relatable.

"So how are you all today?" he asked.

Muted responses.

He went into his spiel and mentioned that he had a PhD in management psychology.

"So, you aren't a labor lawyer?" I asked.

"No."

I asked if I could record the session, and a couple of us did. He made USW out to be a shadowy organization that wanted to take our money and implement a layer of bureaucracy that wouldn't help anyone and would only make it harder for HCL to operate.

"Right now, you can go to your boss directly, but if there's a union, you have to go through them. If you bring in a third party, things get complicated."

"We asked USW for help," I said. "Quit calling them a third party. Our coworkers make up the union."

He backed off, looking offended. It was performative. To me, foremost, he was an actor. I don't know if he believed what he was saying. But we were stuck in this room with him.

He handed out a graph that was supposed to represent USW contracts and strike ratios, but the axes weren't labeled, which rendered it meaningless. We shot him down at every turn, talking over him and arguing his points. I might have been overly confrontational, but we didn't give him an inch, and no one took him seriously.

Kate was well prepared and asked about other companies Vanetti had consulted for. One supporter walked out, visibly upset.

In the next session, someone asked him if his PhD was in union busting.

"I'm just trying to make a living," he responded.

"So are we."

Another meeting got so rowdy, they stopped it early. When it ended, the anti-union folks stayed late to talk to Vanetti in a smaller group.

That night, the organizing committee piled into our meeting room to debrief, feeling victorious. We were energized, and we had rallied the base and brought on more activists. Kate went from a three to a one and started talking to a few soft threes and fours who she knew from the early days. More importantly, she proved to be a huge asset by playing undecided for as long as possible. This wasted the counter-organizers' time, and it helped us gain insight into what they were telling people to try to get them to flip. She kept her promise and was all in after the petition went up. She helped keep Carolyn and Alice on board too. Catalogs stuck together.

September 2019
EMAILS FROM THE BOSS

Later that week, Jeffrey started emailing us regularly with boilerplate anti-union messaging. These continued until the day before the election, alternating between the carrot and the stick. HCL didn't seem to have a deeper content strategy. There was no shape to the story they were trying to tell.

They took lines from the NLRA out of context to make them look even worse than they were to prove that we might not even get a contract during negotiations. Conversely, they said that if we won concessions from HCL that were too great, Google could terminate our contract.

At the same time, they offered doom and gloom scenarios, describing the union as an all-powerful outside entity that would take away our personal choice and individual freedom, all while management was simultaneously downplaying the union's effectiveness. Mixed messages like these clogged our inboxes every day.

A few points they told us:

- There were numerous vendor companies lined up to bid for our current metadata project if Google chose to terminate its contract with HCL.
- The outcome of this situation was independent of HCL's contract with Google.
- We currently had the right to negotiate individually with our employer; unionizing took away our individual right to negotiate with HCL.
- Employees of the union, not likely actual coworkers of ours, would negotiate with HCL on our behalf.

- If the majority of people who showed up to vote voted to unionize, we would all be represented by the union and affected by a future contract. It was not optional.

They combined the scare tactics involving Google firing everyone or not renewing our contract with the classic lines about having an open-door policy and the union being a third party. No one who had worked there had ever been able to negotiate more money with the boss. A few people had received extra vacation in lieu of a raise or a promotion, but that depended on their supervisor. It was an open secret.

The emails from management were easy enough to ignore. If anything, they were so classist and poorly written, they helped us. More worrisome were the personal stories written by our anti-union coworkers. Even if they didn't influence people to vote no, it hurt that our coworkers were willing to put their names on these anti-union messages. It was hard not to take their stories personally. They were mostly scared they'd lose their jobs. The stories ranged in tone from genuine concern to pseudo-intellectual drivel. A few anti-union folks also made YouTube videos, but they didn't get much traction. The videos were slide shows narrated by one of the counter-organizers, and they presented the anti-union folks as the adults in the room who were smart enough to see that there was no point in trying to make things better. None of the emails included any actual negative experience with unions. Everything was pulled from right-wing, anti-labor talking points, with a personal plea to not upset the status quo. It was disappointing to see them side with management instead of with their coworkers who were fighting for them.

The stories and videos included the following points:

- Google isn't covered by retaliation laws, and they won't renew our contract.
- At my previous [non-union] job, all the contractors were fired and replaced by cheaper labor. If we get wage increases, Google will do the same thing to us.
- We wouldn't get wage increases so it wasn't worth the effort. It's too big of a risk.
- HCL was bad, but at least we had health insurance.
- HCL was actually good. I had a worse job once. We should be thankful.
- I want things to go back to the way they were. This is too confrontational. I'm upset.

In addition to the rugged individualist narrative Americans are bombarded with, the lack of a real social safety net provides a more tangible barrier to unionizing in the US. Tying health care to employment makes workers less likely to risk getting fired for organizing. When employers don't face real consequences for retaliation, they are incentivized to bend the truth or lie. They can hold employees' insurance over their heads or fire someone as an example to other workers. It happens all the time.

Where management messed up was sending these messages to our personal email addresses, which they shouldn't have had access to. Whoever pulled the email addresses from the company records didn't scrape the information properly, and a bunch of emergency contacts received anti-union emails. People weren't happy about it.

September 2019
THE ANTI-UNION COMMITTEE

We continued to get personal stories from other coworkers about why we shouldn't unionize. They were mostly more fear-driven appeals to the status quo that begged us not to do anything drastic that would threaten our relatively good jobs.

Some of these fears weren't unreasonable. We'd made a calculated risk that it would be more of a hassle for Google to cancel the contract in the current climate of bad publicity than it would to be for them to pay us slightly more. We didn't even know how the contract was structured. It was possible HCL would make slightly less on a new contract and Google's bottom line wouldn't be affected. And there was nothing preventing Google from ending our contract as it was.

Every day I went into work and felt like I was a whisper away from puking. I stopped listening to music so I could hear people approaching and catch snippets of side conversations. In the meetings, union supporters hammered Vanetti and tried to wear him down. One worker got him to admit that we had nothing to lose by unionizing and that the company shouldn't have let it get to this point. I doubt that's what he said in private to the anti-union group, but we recorded it. He made an effort to seem objective, and the workforce continued to meet with him until the day before the election.

HCL sent letters to our home addresses, and supervisors pulled us into one-on-one meetings to try and assess where we stood. They'd never had answers for us when we needed work-related information, but they were suddenly on-call and available if we wanted to talk about any misgivings we might have had about the union.

Our workplace was divided. The anti-union committee spent three weeks working out in the open, on the clock, and with the aid of management. They were trying to befriend people they thought were undecided, and they criticized the union in front of the known organizers. At times it got personal, but we couldn't respond at work. They worked frantically to figure out who they could sway. Meanwhile, Vanetti held his meetings, and management pulled workers into one-on-ones. Mark emerged as one of the leaders of the anti-union organizers. He and others were telling people that Jeffrey said HCL would move the office to Minnesota if we unionized.

Most of the threats and fears were related to the possibility that HCL would break the law and illegally shut down our team if we unionized. If we had better labor laws in America, this narrative wouldn't have been able to gain traction. If we had card check recognition, we would have been able to start bargaining as soon as more than 50 percent of us signed cards. It would have taken us about two days to get that many, and we wouldn't have had to deal with Vanetti or any of the counter-organizing bullshit.

Despite decades of antiworker propaganda and corporate-friendly legislation, though, union popularity is on the rise. Unfortunately, it is harder than ever to form a traditional union.

September 2019
BIRTHDAYS

Gracie turned twelve in early September, right around the time we filed for our election. She wanted to go to her grandmother's in Evans City for her actual birthday and then come back for a party with her friends. I worried her friends were setting unrealistic expectations about what birthdays were supposed to look like. Gracie's party consisted of her and two friends going to the mall for the day. I dropped them off with money for a movie and junk food, and I played pinball for a couple hours while they walked around.

The following day, she and I had a cake at home and a few gifts. She requested steak for her birthday dinner, which I gladly cooked, and we ate together in front of the TV.

"When the downstairs neighbor moves out and we live in the whole house, you can have sleepovers," I said. I had no idea when I would be able to afford to pay the entire mortgage, but our plan was to move into the whole house as soon as possible.

"Why can't I now?"

"It's too messy."

"I'll clean my room."

"We don't have the space," I said. I felt guilty for depriving my daughter of such a normal childhood experience. I also felt like I wasn't holding up my end of the bargain with the families who occasionally had Gracie stay at their house overnight. In my mind, they were silently judging me. If not judging, then pitying, which was definitely worse.

Even after cleaning thoroughly, our apartment was still cluttered. For most of Gracie's birthdays, I did what I could on a budget, and in a public place. Cake and pizza in the park was easy enough. One year we went to the trampoline place way out in the suburbs with a couple friends.

"Everyone else has sleepovers," Gracie said.

"You can do anything you want besides having people over. Except Dave & Buster's."

The Dave & Buster's party was brutal. I'd spent almost $500 on tokens for her and her friends. It was the last birthday Gracie spent with her mom, who was on supervised visitation at the time. She and the social worker showed up at the party, and Jane was helpful and gracious. She really seemed like she was turning it around that day. She even chipped in and paid for half the food. That was two years ago. Gracie's birthdays before then had almost always been the same kind of scene—the kids having fun while Jane passive aggressively bad-mouthed me to other parents and tried to provoke me into doing something stupid.

Right before Gracie's first birthday, Jane and I got into a shouting match. She accused me of being a cheap piece of shit for buying used gifts. I figured Gracie wasn't going to remember anything about the day anyway. But I eventually relented, putting a bunch of stuffed animals and books on my credit card.

Then there was the Chuck E. Cheese incident when Gracie was four or five. Jane threw a party for her and didn't invite me. Then she tried to tell the court I had skipped Gracie's birthday. When I went to pick Gracie up at the appointed time, no one was home. I called Jane and she didn't answer. I got a hold of Jane's sister eventually and she told me where they were. So I drove across town to Chuck E. Cheese, where Gracie had insisted she have her party. The whole family was there. Jane said her phone had died.

"I'm glad I didn't miss it," I said.

"Me too," she said. "Gracie, your dad came!"

"Hi, Dad, love you, Dad," she said, and she scurried back out into the ball pit or onto the jungle gym.

Eventually, we started alternating birthdays, and the other parent would throw Gracie a second party on a different day. As a single parent, I've grown less attached to particular dates in general. Christmas was sometimes on December 27, and Thanksgiving dinners sometimes took place the Saturday following the official holiday. It was more about the time together with Gracie.

For her twelfth birthday, I bought her an iPhone and some video games. Presents were getting more expensive. When her mom was alive, I'd overspend to "win" the present competition, but this always led me to feeling bad for Jane, who was hurting for money even more than I was because she was spending it on heroin. There were years where she bought awkward gifts that weren't age appropriate. She would take Gracie to get a makeover as a toddler, or she'd buy her stuffed animals or Barbies after Gracie had stopped playing with them. I would put a video game system on my credit card and call it a day.

The year of the Dave & Buster's party, Jane was clean. She got Gracie a new bike that she kept up at her grandma's house in Evans City. Gracie would come home and tell me how much fun she had riding up and down the big hill by their house and exploring the town with her friends in Butler. That was a great gift.

September 2019
PRESS TRAINING AND THE GOOGLE ACTIVISTS

My press training with USW took place over a phone call I took in my car during my lunch break. The media around the union drive was intense. I'd done a few interviews about my book, and I'd occasionally gone on podcasts to talk about writing, but it was a whole different thing when it felt like the fate of the union hung in the balance. As one of the first organizers, I started doing a ton of interviews. I was so shot from the months of stressing out about getting cards in that I could barely put two sentences together. But I learned a few easy rhetorical tricks to dodge unwanted questions and stay on message. It got easier the more I did it. After a few bumpy appearances, they smoothed out.

Before an interview, I would research the journalist in question and make sure they weren't going to try to take something I said out of context and make us look bad. Thankfully, that never happened. If anything, they made me sound much more succinct by pulling the good quotes out of some of the meandering, sleep-deprived interviews I gave. The podcasts and panels were tougher. I think I bombed a couple of early appearances due to nerves, but I mostly held my own. Sometimes, I'd find myself talking in my polite phone voice and had to snap out of it. This usually happened when I couldn't think of anything to say.

After the USW press release stating that we were seeking a union election went public, a handful of us on the organizing committee were invited to group chats on various encrypted apps with the people who'd organized the 2018 walkout and were still in the midst of their own fight with Google. That day I ran into

Paige at lunch, and we marveled at the network we were now part of. There were a handful of TVCs, but it was mostly full-time employees. They were ecstatic for us. We had a Zoom call with one woman who'd been an activist for years, and she was so happy for us she was weeping. They all pledged to help.

As word spread and the media outlets ran with it, we made more connections with other tech workers and nonindustrial workers who were organizing. We were invited to talk to workers all over the country about how and why we wanted to unionize. I met the Kickstarter organizers who had been fired during their own union drive and dozens of other tech worker activists.

I met one of the women who had started the Google walkout, and a few other San Francisco organizers who tried to take over after she was fired. There was a strong contingent of Google workers who continued organizing behind the scenes. One of our biggest supporters was a Communist, a full-time employee who worked in another office. That worker taught me a lot and has been an incredible mentor and ally. For example, they intervened when another full-time employee tried to co-opt our union drive to further his own message. It was a delicate situation. We needed everyone's help, but there was clearly a disconnect in strategy between some of the organizers. Some asked us what we needed and listened to us. Others told us what they thought we should do and tried to impose a plan of action. Another Alphabet worker harassed Eric Vanetti so badly, Vanetti deleted his Twitter account.

The vast majority of former and current Google employees were supportive. They were vocal in ways we couldn't be without risking our jobs. They used their positions of privilege to their advantage and did a ton of organizing on our behalf, helping us build a strong coalition of tech activists. Their vocal public support helped ease our fears that Google would break their contract with HCL or send us to another office. They found the documentation that stated Google could not end a contract due to employees unionizing and pushed Google to make a public neutrality statement.

A group of Pittsburgh full-time employees visibly showed their support by wearing buttons on-site. As I got to know some of them, I learned most of them thought we made twice as much as we did. They had no idea we were being treated like second-class citizens to the extent that we were. They were self-aware, recognizing how good they had it in many ways, and they supported us and other Alphabet workers who were in even more precarious situations.

I constantly doubted my parenting decisions, but I never doubted I was doing the right thing by unionizing my workplace. At times, that led to me spending too much time and energy on union stuff at the expense of parenting. It wasn't that much of a time commitment, but the mental energy and stress was heavy. It was hard to relax and not think about it. Doing the work was easy to justify. I was trying to improve material conditions for everyone at work, including myself, which would still benefit my daughter. If we were successful, it would help all workers, especially underpaid contractors. I tried to thread the needle as best as I could, but I enjoyed talking about unions and work and supporting other places who wanted to start unions. I said yes to almost every interview, every organizing call, every podcast, every panel discussion, and I met amazing people because of it. But after a while, it became clear I was putting in an unsustainable amount of work. I'd have to start saying no eventually.

September 2019
TWO WEEKS UNTIL THE ELECTION

Abby had worked on Google Shopping since before HCL. She'd been there long enough to see the free parking and floating holidays disappear. She was outspoken in her distrust of management. She was on board from the beginning of the union drive, and she sat in on a number of the anti-union sessions to take notes and report back if any of the talking points seemed to be gaining traction.

One afternoon, she saw all the anti-union folks walk through the library and out onto the balcony overlooking Target. She texted me, and I headed toward the parking garage on the other side of Penn. I ran into Paige on the way, and we walked up to the rooftop of the garage and stood there, trying to make out who was meeting. It was a bright, sunny day. A million different ideas, all of them bad, streamed through our heads as we strained to ID our coworkers. They were definitely on the clock. Will was a floor above them, trying to listen in on their meeting.

It was ridiculous, really, trying to spy on them from the other side of the street. We couldn't tell anything other than that they were all sitting together on the patio. Will and Abby could maybe hear snippets, but that didn't matter. There was no point in watching them. But we knew they hated to feel like they were being surveilled, so we watched anyway.

"It's amazing they were able to just mosey out there during work hours," Paige said.

"I wish I could tell what's going on," I said.

We stood out in the heat, too far away to discern anything, watching our coworkers plot against us. It was absurd, but we stayed out there watching until we couldn't justify the sweat any longer and went back inside.

I got into a new work routine during Gracie's fall softball season. I'd stay late on days when Gracie didn't have games so I could leave early on Wednesdays and Fridays when she did. This was particularly important when she played 6:00 p.m. games in the West End, Brookline, or Carrick. On those days, we had to get across town and over the bridge during rush hour. It was terrible. The rides home weren't as bad, at least, and the early games allowed me to play on my adult league team at 9:00 p.m.

When Jane used to come to Gracie's softball games, she lingered near the bench and talked to Gracie in the middle of an inning in a way that made me uncomfortable as a coach and co-parent. It went beyond passing a drink over the fence or handing her a sweatshirt if it was chilly outside; Jane would stand there and start full conversations about random shit. She'd end by telling Gracie to put on sunscreen or to drink more water when I had already made sure she was hydrated and sunscreened. It was as if Jane didn't see it as an organized activity for the kids but something she needed to help Gracie with, like her daughter was still a toddler.

Despite our custody agreement, Jane still treated everything that Gracie and I did together as if I were an absentee dad who picked her up every other Christmas and had finally decided to participate in my child's life. It was performative and intentionally hurtful at first, but after a while, it was just annoying and sad. There were games where she was totally glazed over and not making sense. She would talk in this affected, almost childlike voice when she would ask Gracie if she had remembered her hat or something, and I'd just tell her we were OK. When Gracie played the outfield, Jane would walk down the foul line and talk to her during the game, which was not typically something parents did. Gracie was always happy to see her mom, but she sometimes looked uncomfortable and unsure of how to act. It was like she was being pulled in two directions. Maybe I was seeing something that wasn't there, but it seemed like Gracie was especially concerned about hurting her mom's feelings by not talking to her during games. Jane pulling Gracie away from the game was not surprising to me. Softball was something Gracie and I did together, and naturally, Jane wanted to wedge herself between us.

As politely as I could while still being direct, I'd ask Jane to wait until after the game and not to come to the bench. She was always high at that point—when I had primary custody but before Jane was on supervised visitation. When she was on supervised visitation and was getting random weekly drug tests, she was less intrusive. She seemed defeated. Sometimes, when Gracie only got to see her mom one day a week with a social worker, I'd break the custody agreement and let Jane take her for ice cream and hang out for a couple hours after games before dropping her off at home. I'd usually try to nap during that time. Gracie missed her mom and I was always tired.

I felt relieved whenever we had an away game and the bleachers were far from the dugout or separated by a fence or something that didn't allow Jane to easily bug Gracie. These basic social norms of youth league sports were important to me, and maybe I was overly concerned, but it made me uncomfortable and a little embarrassed for her. I'd text Jane before the game to not bug Gracie when she was playing, and she'd flip out and say she wasn't doing anything wrong. I'd explain that parents weren't allowed on the bench. She either didn't care or didn't understand. Sometimes she'd wander around the field, beyond the center field fence at Mellon Park, around the ice cream truck in the parking lot, and into the shade by the tennis courts. Like she couldn't see the game, even though it was right in front of her.

I used to be thankful that Jane worked a lot of weekends. I was always happy when she couldn't make the games. Gracie played better and didn't seem upset about her mom's absence. She paid more attention to her teammates and the game. A lot of the time, Jane would skip games she planned on coming to, saying she had picked up an extra shift.

But now, Gracie didn't have a mom to cheer her on from the stands anymore, and I felt like a terrible, petty man for feeling unsettled by her behavior as a spectator.

Gracie improved a lot just by practicing and playing more. I hoped that would be a lesson about hard work and practice that she could use to get better at things she wasn't naturally inclined toward, like math problems. Those are the lessons I learned playing baseball, anyway. I was OK, but not a natural, and I had to work hard. This is how I am as a writer, too, and I guess it's how I am as an organizer. Often, I've thought that if I didn't still play baseball and write, and if I wasn't trying to organize my workplace, I could have probably dedicated myself to something more sensible and given Gracie more stability. Sometimes I wondered if I was wasting my time, spreading myself even thinner instead of finding a better job that would have allowed me to pay for her to go to college. But maybe I'd be even more miserable that way. I don't know.

Maybe the whole union effort was part of a midlife crisis that I justified because I was helping people. Everything about the capitalist society I live in wants me to internalize blame and second-guess myself for trying to make art, have a social life, exercise via sports, and raise my kid instead of spending all my time working until I can't anymore. It was so hard for me not to blame myself for doing my best.

But my constant second-guessing never helped much. I was always fighting a battle to stay present, except when I was watching Gracie play softball and, to a lesser extent, when I had a baseball game. For a couple hours, there was only the game before everything else came rushing back.

September 2019
OUR UNION PRESENTATION AT THE LIBRARY

Gracie had practice so I got to the presentation late. I grabbed a seat in the dimly lit back room of the library and started worrying again. I hadn't changed after practice, and I smelled like baseball, dirt, and sweat. Sam, Paige, and Jenny put together the presentation. They alternated narrating as the show progressed. Will sat up front and helped answer questions. It was better if Damon did as little talking as possible, stepping in only when there was something super technical we couldn't answer.

The anti-union committee sat in a row right in the middle and scoffed loudly throughout the presentation. They all held sheets of paper, presumably with questions on them they were supposed to ask. When the lights came on after our slideshow, the anti-union pageantry began.

Someone tried to record the question-and-answer session, but Kate shut them down.

"I'm not comfortable with you recording this," she said, standing up. "I think others would prefer management not to have any insight into their leanings."

There was some disagreement, but the guy turned off his phone.

Mark stood up and spoke, "I've worked here for five years, and I wasn't worried about my job until last week. Google will fire us for cheaper labor. I was fired from a contractor job in Brazil, working for a Fortune 500 company. They found cheaper labor and fired us all. I demand we withdraw the petition for an election!"

The ten or so anti-union folks all stood up and clapped.

Wendy was concerned about what would happen in

negotiations. Damon explained it, and she said, "There's no way I'll even be on that committee."

The anti-union workers got more and more excited with each predictable question. They shouted down our coworkers when they calmly tried to answer bad faith questions.

To finish the session, Jack stood up.

"It says your organization stands for unity, equality, and equity," he said. "Why should we pay dues to an organization we don't like, for protections we didn't ask for?"

There was another demand for petition withdrawal and more clapping from a third of the attendees.

"We follow the law," Damon said.

Paige thanked everyone for coming.

On his way out, Jack yelled, "This fight isn't over!" and stormed out.

Paige thanked him again and waved politely. We would shout at Vanetti but not the people we worked with every day.

After the presentation, I helped put the chairs back and talked to Bobby. He said Darren had taken a new job and was working at the new place remotely from Google's office while he stayed on the payroll and helped with the anti-union effort.

"At least he's leaving soon," I said.

"I told him I was undecided and didn't want to talk about it," Bobby said before he left. The rest of us went down the street to Kelly's Bar. The meeting had been stressful, but we felt good about it. We were happy it was over. We bought rounds for our coworkers who took the brunt of the abuse during the meeting and responded calmly with a smile. They were amazing. I couldn't have done it.

The next day, Jack apologized to Paige on the corporate chat so there would be a record of it. The anti-union emails kept coming. The sessions with Vanetti in the Spring Hill Suites continued. Management was panicking, and they ramped up their efforts to scare us as we got closer to the election.

Will: Hey Ben, what if we leaked fake information to them so they wouldn't know who to organize?

Me: What if they know it's fake?

Will: I thought of that. We could test it out. If we marked, like, Bruce as a maybe, we could see if they try to recruit him.

Me: I don't know how we'd leak a fake spreadsheet.

Will: We could leave it out where they'd find it.

Me: Seems like more trouble than it's worth.

I wasn't sure how seriously to take Will's ideas, most of which I agreed with in theory. Some, like his suggestion that we frame anti-union workers and managers for intellectual property theft, were clearly a way to blow off steam, but I worried he'd take things too far.

One of my favorite organizing strategies is from the United Mine Workers (UMW) leading up to the 1913 strikes in Colorado. UMW organizers worked in pairs; one would get a job at the mine and cozy up to management, while the other signed up workers for the union. If someone didn't sign a union card, the first organizer would tell management they were pro-union and get them fired. Then the second organizer would send a pro-union worker to take his place.

The stakes were certainly much higher then. My coworkers and I were not in danger of being shot by private detectives or dying in a collapsed coal mine. It is inspiring to read about the lengths early twentieth-century workers went to in order to fight for better working conditions. Workers died to give us the right to unionize at our office job for better pay and fair treatment. As far as Will's hypothetical schemes went, I had no moral problem with bending the rules to beat management. They certainly weren't playing fair. But as a practical concern, it was important we maintain the high ground and not risk any action that could be considered intimidation. We also didn't want to get anyone fired. So we stayed aboveboard in everything we did.

September 2019
ONE WEEK UNTIL THE ELECTION

Dear Coworkers,

We know you've been getting a lot of emails recently, and we appreciate you taking the time to read ours.

This has been an eventful week.

It would take pages and pages to respond to all the baseless claims we've heard, but we would like to counter a few points.

The truth is a union is entirely democratic. We are the union. If we win the election, we will be our own local, an independently run chapter of the USW. We will have help negotiating our contracts, but HCL workers will be there at the bargaining table. Everyone has a chance to be elected to the bargaining committee. Nothing goes into a contract that is not approved by a majority of our local members.

Eric and Jeffrey make it sound like the USW could call a strike at any time and we would just have to go. While the International's president *does* have to sign off on any strike to ensure that it is legal, the International does not call for strikes that members have not approved via a strike authorization vote. Strikes don't work without support from the workers. An unwanted strike would be a disastrous failure and would almost certainly lead to a union decertification campaign.

In an email, Jeffrey asserted that there have been twelve USW strikes "in Pennsylvania alone" in the last decade. There are approximately 110 USW locals in Pennsylvania. An average of 1.2 strikes per year, divided by 110, means that each union local went on strike an average of 0.01 times per year, or put differently, about once every one hundred years.

Most locals set the vote threshold for approving a strike very high, often as high as 90 percent needing to vote yes, in order to ensure that a strike genuinely has the support of the membership and will actually be effective. Across the board, approximately 98 percent of USW contracts are resolved without a strike. Strikes don't happen unless every potential avenue of negotiation has been explored and there are no other viable options.

No one has to strike if they don't want to. Forcing someone to strike is illegal. The USW International does not fine members for not striking, period. A small number of locals, by a vote of their own members, have amended their bylaws to allow for this. Whether our local does that or not would be up to us.

Some unions do pay officers, but this typically only happens in unions with much larger membership numbers than ours would have. In some cases, officers (usually just the president) are granted a few extra hours of PTO in order to tend to union matters. Most officers work for free. Who becomes an officer is something that is voted on and varies with each local. Eric and Jeffrey would have us believe that the organizing committee is only doing this so that we can become officers and get preferential treatment.

It is likely that Eric and Jeffrey will push even harder in the next few days, leading up to some kind of desperate "bombshell" right before the election.

Remember, it is illegal for management to ask how you plan on voting. You do not have to see or talk to Eric if doing so makes you uncomfortable. He has specifically not been allowed into our office, and HCL cannot force us to leave the office for meetings.

It is important to keep in mind that we are doing this work for everyone in the bargaining unit. We all deserve better. It is natural to have a strong response to potential changes in the workplace. It is completely understandable to seek out the familiar in times of conflict and uncertainty. We hope to win our election, unite our workplace, and give everyone a voice. We are not working against our coworkers who are concerned about the union. We are fighting to get our fair share from a multi-billion-dollar company, a company that claims slim margins when we ask for better salaries but pays a professional union buster to pit us against each other. More than two-thirds of us signed cards because we want a better workplace for everyone.

Thank you for your continued trust and support.

In solidarity,
HCL Organizing Committee

September 2019
PICK UP YOUR CLOTHES

I had cleaned Gracie's room thoroughly, organizing it so she could theoretically have a clean place to do schoolwork. The school year was barely a month old, and she was already falling behind. We spent another weekend getting her caught up with all of her classes.

"The quotation marks go outside everything," I said. "Make sure you do it like this going forward."

"OK. I know. I'm sorry."

We'd been working for several hours, bouncing between the multiple content management systems, the special math program, and the history dashboard. Gracie's clothes were still scattered across the floor in piles, and there was an open bag of chips on her bed and crumbs on the sheets.

I didn't really get on her case too much about eating junk food because I cooked relatively healthily, but the clothes and the crumbs and the homework that hadn't been done for weeks, again, was getting to me. I couldn't get away from stress at home or in the office. I wanted to shut down and check out, and it wasn't fair to my kid. I told myself I'd been trying to help her establish independence and good study habits without me looking over her shoulder. But it was also easier for me to provide her with less structure. I didn't have the energy to sit with her every day and go through everything. I made myself available, I told her to ask me for help, but she rarely did until I got an email from her teachers. I urged her to do the work even if she didn't know the answers.

"Especially writing assignments. People don't even know if they can be graded. Just meet the length requirements, let me proofread it, and you'll get a B."

"OK. I will. I'm sorry."

"Can you pick up your clothes, please?" I asked when we finished homework for the day.

An hour later, I asked again, "Please pick up your clothes."

The third time, I yelled.

"Pick up your fucking clothes and get the food out of your bed or we're going to get mice. Do you want mice?"

Gracie cried. I felt like shit. I apologized for getting mad, and I helped her clean her room. "Seriously though, we don't want mice," I said.

I'd fall into cycles like that where I'd ask and ask and ask, and then I'd lose my cool and yell, and it would feel awful. I don't know why that was my default solution. I'd never yelled at her like that before, and she'd never previously lied about doing homework. I felt like we were wading into something difficult that I wasn't equipped to handle. There had been other moments—like the time she somehow ran up a $300 cell phone bill and I yelled at her or the time she almost stabbed me in the eye with a pencil and I swatted it out of her hand reflexively—but those were different. I was worried I was turning into a shitty dad who yelled all the time at a kid who resented him.

I tried bribery, punishment, more structure, less structure—nothing seemed to work. Grounding her was counterproductive because I didn't want her in the house all weekend. Sometimes, I'd sit and watch her do homework only to find out she hadn't handed in the completed assignment. Sixth grade had been a wash and seventh wasn't going well either. Hopefully, we could pull it together for eighth grade. If not, high school always offered the promise of a fresh start.

September 2019
THE DAY BEFORE THE ELECTION

I came into work the day before the election and saw Austin dressed as Uncle Sam. The day before, he had called the organizers something like "bourgeois social justice warrior communists." Now, he was dressed like a WWII propaganda character. The office had taken on a on a dreamlike quality since we filed cards, and today it was peak surrealism. The anti-union folks were still hustling to try to change minds and being overly nice to everyone. We were wearing our union pins and putting on a positive face and making sure all the pro-union workers had a plan to get to the voting location. I don't think anyone got much work done that day.

Exactly twenty-five hours before the election, management hurried us into one last mandatory meeting, allegedly to go over election logistics. I asked if I could skip it, but Donald said no. We all walked across the sky bridge and into a conference room I'd never been in before. Once we were seated, Tom and Jeffrey gushed about how great HCL was and how they were going to expand our benefits in line with Google's policy requirements in 2022. Jeffrey's face was bright red, and he stumbled through his speech. A few of the anti-union workers asked layup questions and nodded enthusiastically at Tom and Jeffrey's talking points.

I raised my hand.

"I thought this was going to be about logistics," I said. "I'd like to leave, please."

Jeffrey said that was fine. I got up, and Carl and Kate followed. A few people stayed behind and took notes. During the meeting, Jeffrey told everyone that we would lose our flexible scheduling if we unionized. We filed a charge with the NLRB.

We wandered outside and met up with Bruce. His team had been pulled into a meeting with Ernst that had been just as bad as ours. The counter-organizers were getting bolder. Hopefully, I thought, that was a sign of desperation. I was worried folks would get scared right before they went in to vote and would decide to keep things the same. After all, there were worse jobs. We could be making $10 an hour, working in a basement call center, or doing any number of things for less money. I played out a million different worst-case scenarios in my head.

After we went back inside, a cake arrived from the organizers in San Francisco. It was delicious and helped cut the tension. At lunch that day, more and more people wore pins. I sat in the cafeteria with my laptop alongside my coworkers, commiserating and trying to relax.

———————

I stayed late that night so I could leave early the next day. There was no way I could work a full day after the election. I'd either want to celebrate, or I'd be crushed and want to go home. Around seven, I was alone in the office with my teammates. Someone found a union pin attached to a Post-it note. The note read: "FREE if you want to pay dues, not get raises, and risk everything. Yay! (:"

This was pretty minor as far as the discourse from the anti-union folks went, but it struck a chord. More than anything, it was the timing and tone that pissed me off. It was the same kind of infantilizing bullshit I heard when Donald told us, "this is just how corporations work," when HR fucked up our vacation hours, or when Nancy told us to take on a side gig during the holidays. I hated being condescended to, especially in this context. Like if I just thought about it, I'd realize we were making a mistake. We had worked hard, and we'd been honest about the calculated risk we were taking. We weren't starting a book club. Unless we had drastically miscalculated everything, we were going to win, and we knew the victory would be the result of the strength of the positive campaign we had run. If we lost, at least we tried.

I picked Gracie up from practice at the field behind the Boys and Girls Club and made the short drive home. I ordered pizza, and we watched TV for a while. She said she didn't have any homework, and I believed her.

"I won't be as busy soon," I said. "We have our election tomorrow. Hopefully, we win."

"Why would you lose?"

We were sitting on the couch. Gracie was still in her sweats from practice.

"Because our bosses' boss spent a lot of money to try and scare us."

"That's dumb."

"I know," I said. I grabbed a can of water from the fridge. "I'll be back soon. This will be my last meeting for a while."

"OK, I love you."

Gracie was fine to stay home alone by herself for an hour. On my way out the door, I felt a longing for a time where I would have had to take her with me. I was looking forward to having my evenings free to watch fall ball games and spend more time with her. If we won, the committee would have some work to do before bargaining, but it seemed mostly administrative.

After weeks of saying yes to everything and becoming one of the organizing committee's de facto spokespeople, I wanted to take on a smaller role and stay in the background. I assumed the toughest part was behind us. I didn't like planning for anything after the election because I thought it would jinx us.

That night, I went to the coworking space for one of our last organizing committee meetings.

"There's probably going to be a zero-hour surprise," Damon said. "But if you've been inoculating, it won't matter what it is."

We went over our Get Out the Vote plan. We text-banked, emailed, and called people on our teams. We only focused on solid supporters and maybes. We made plans to go out after the election, win or lose. When I got home, Gracie and I played Mario Bros. for an hour before she went to bed. That night, I slept well for the first time in months.

September 2019
ELECTION DAY

I sat around and looked busy for a few hours, then went to vote. HCL offered van rides to the library where the vote was held, but it was a nice day and we were on the clock, so most of us walked. I went early with Paige and some of her teammates. On the way down, as we crossed the street near the Target, I had a feeling we were going to win, but I didn't let myself completely stop worrying.

Still, seeing everyone out on the street made the possibility of victory easier to digest than looking at numbers in a spreadsheet. We were all upbeat and ready to vote. The anti-union folks stared at us from across Penn Avenue.

Once I got inside, I studied the paper ballot and made sure I marked the correct box, then I folded it and put it in with the others. There were still a few hours to kill before the count. I stood outside, a few yards beyond the legal distance from the voting site with Sam and a bunch of supporters from the community and USW who had picketed on our behalf. I'd switched to vaping since my surgery, but that day, I stole a cigarette from my buddy Rich who had come down to support us. We talked about the preemptive anti-union presentation we got from HR during training when we worked at the steakhouse at the casino, and how shitty that job was.

"They had Red Bull on the soda gun, though," Rich said, and we laughed.

After a couple interviews with the local news, Sam and I went back inside the library for the vote count. We were joined by a group of pro-union workers who had taken the afternoon off and walked down from the office. Will and Mark sat on either side of the NLRB representative in charge of the ballot box. The Ogletree Deakins lawyers were also there, along with Jeffrey, a couple other bosses, and our USW lawyer. The lady from the NLRB counted each ballot by hand, putting the yeses in one pile and the nos in another. She was visibly happy when she counted the yeses.

Mark and Will kept count along with her, along with several of us sitting across the table from them. It was close for the first twenty votes or so, and then there was a wave of yeses. Once we hit forty yes votes, we knew we had it no matter how the challenge ballots went. I felt deep relief and pure joy. I hugged Alice and cried while the NLRB lady finished the count.

We won forty-nine to twenty-four. We had held our supermajority from wire to wire, even though a handful of supporters were on vacation that week and others had left the company entirely before the vote. We had agreed that if we won, we wouldn't make a huge scene in the room, so we silently hugged and patted each other on the back. Then Abby went outside and told the library workers we'd won. They had recently won their union election with USW too, and when they heard the news, a roar came up from the workers downstairs. Damon almost passed out on the steps on his way out the door. Outside, there were more reporters. We talked to the media and then went back to the bar.

But we didn't have much time to celebrate.

EPILOGUE

March 2020
REMOTE WORK

When the COVID pandemic hit, I felt incredibly lucky to be sent home with my laptop to work from my living room. People lost their jobs and their housing. Hundreds of thousands of people died, many of whom were forced to go to work because our government wouldn't pay everyone to stay home, even when the clear answer was to promote togetherness and communal solutions to the pandemic. But it was portrayed by the media as a war between left and right over personal choice. Individuals were blamed for our country's ongoing inability to stop COVID instead of the systemic problems that prevented America from implementing real solutions. If we had universal health care, and a real social safety net, fewer people would have died. We got a public-private mess full of half measures instead of ongoing stimulus payments and cohesive vaccine and mask distribution. Democrats blamed individuals who didn't wear masks, and Republicans claimed vaccines and mask mandates infringed on our freedom.

It took a global pandemic for Google and HCL management to admit they had always been able to let us work from home. The days dragged by, and the weeks ran together. The first weeks, when we were unsure of the extent of the virus and we were still wiping down our groceries, have taken on a dreamlike quality. I took on another side job as a technical writer. I worked on my own writing or freelance work during the day and put off my HCL tasks until late at night. I still wasn't as productive as I would have liked, and I couldn't get out of that mindset, even while working around sixty hours a week between a handful of jobs and my own creative work. I'd work until I'd hit my quota and then go for a walk, late at night when Gracie was asleep upstairs and the streets were empty and quiet. I was perpetually worried about what was going on outside and across the country. All the external pressure piled up, and it was suffocating.

As we started bargaining our first contract, there seemed to be an upswing in organizing and renewed interest in workers' rights. I got more involved again too. I sat on panels and agreed to do podcasts. I did more organizing calls with other workplaces. I signed up for dozens of online union training courses. I read as many labor books as I could, history books, and organizing stories. I joined a nonprofit organization that advocates for temp workers and tries to get legislation passed to ensure equal pay for equal work. Our goals include putting an end to perma-temping and making third-party employment less profitable for companies. With the help of my second job, I paid off my car and my credit card debt.

I felt guilty for improving my situation at a time when so many people were dying or being forced to choose between going to work and risking getting sick or missing shifts and not paying rent. Even though that line of thinking was counterproductive, it was hard to shake. I felt guilty about my anxiety over work and the guilt made my anxiety worse. I started having trouble sleeping again.

One night, I fell asleep on the couch in front of the TV and woke up to the couple across the street screaming at each other. Still groggy, I watched out the window, feeling grateful that I was sober during quarantine. The dad still walked with the kids down the street around 3:00 p.m. and then walked back with them around midnight. I'd see the mom waiting out front, staring at her phone, smoking cigarettes until she got a ride around the block. Recently, I saw her panhandling by the bridge. Her sign said, "Down on my luck. Anything helps." The light was green, and I couldn't stop. They were evicted in December 2021. According to court documents, they owed $12,000 in back rent. I watched from my second-story window as contractors pulled out the old appliances onto their stoop and tossed them on the back of a truck.

July 2021
BARGAINING AND COVID

We started bargaining in November of 2019, and we ratified our contract on July 28, 2021. It took almost two years, but we eventually got it. The contract was good but not life-changing for most of us. The four lowest-paid workers saw significant increases; they went from between $36,000 and $37,500 a year to $43,000. Everyone else got just under a 10 percent raise spread out over three years. We got four more days off, including MLK Jr. Day and Presidents' Day. Management is still trying to get out of Google's mandate that we receive dedicated sick days that are available without accrual. We negotiated a clause that says we can't receive fewer sick days than non-union TVCs. We also negotiated the accrual rate for our PTO.

The transition out of organizing and into bargaining was difficult. It wasn't easy to shift gears after we'd won. We polled the workplace on what they wanted us to prioritize in bargaining, and we were in the room with the lawyers from Ogletree Deakins and Jeffrey and the HR guy who loved our ticket system. We had input on everything. The best thing we did was bring in the most well-respected anti-union people and make sure they got seats on the committee. They joined and helped garner union support from others who'd been on the fence.

We brought up COVID at our bargaining sessions in January 2020, and the company had no contingency plan. Their lawyer said they weren't concerned about it. Full-time employees were sent home first. As time went by, the office gradually grew emptier until we were the only ones left. Other than the pandemic, it was quite pleasant to have all that space to ourselves. It was like when the Google employees went on their ski trips and other team-building outings during the week. It was quiet. We watched TV together in the game room. No one was working too hard.

But it took collective action for us to finally get sent home. Workers with children were allowed to leave. Then, by the end of that same week, everyone was allowed to leave if they wanted to. Eventually, we were all sent home. We were assigned training in Microsoft Teams and told to complete modules that gave us advice on how to work from home. We waited for Google to get our permissions figured out, and then we worked from our living rooms and home offices as if we were all still in Bakery Square.

On Catalogs, while we worked remotely, our metrics became comically strict, and we lost all variety in our task work. We lost the permissions for documentation that we needed to do our jobs well. We no longer had contact with Google engineers, even on collaborative projects.

For almost two years, our bosses stopped hiring in Pittsburgh and replaced workers who quit with new hires in Kraków. We filed over two dozen unfair labor practice charges. We subpoenaed agendas and meeting notes from meetings between our bosses at HCL and Google. Donald didn't attend our daily check-ins. Instead, he sent a middle manager to meet with us. He wouldn't answer basic questions about staffing and our task work. To us, these kinds of actions felt trivial and childish while the country was on fire amidst the biggest protest wave of our lifetimes.

We dropped all the ULP charges in exchange for the ratification of our new collective bargaining agreement. Retaliation continued after we had a contract, but mostly within the legal parameters of the NLRA. I expect we will file several grievances. The transition out of bargaining and into running the union has been rough, but I'm sure we'll get through it.

The Black Lives Matter protests in the wake of the George Floyd murder, combined with a renewed interest in the rights of essential workers, gave me hope that things could change in our country, but nothing really has. Biden has been predictably moderate at a time that calls for bold, progressive policy. When NBA players were on the verge of a meaningful strike during the playoffs, former President Obama recommended they get back to work and start a voting project instead. A moderate Democrat promoting the individual act of voting over collective action like strikes or protests is not surprising. We're supposed to vote and then blame the other individuals who voted against our team or didn't vote at all, instead of building a better team that really has our interests in mind. We are still a country that puts individual "freedom" above the needs of working people. I don't think that will change in my lifetime.

More tech workers are unionizing, though, including workers at Google. The Alphabet Workers Union formed in January 2021, led by some of the same activists who helped us. As of October 2021, they had over 800 members. As a minority union, they don't get to bargain for a contract, but they fight for the most vulnerable workers and for more control over their labor. Kickstarter won their election and is still negotiating. Tech workers at the *New York Times* organized and became one of the biggest tech unions in history to win an NLRB election.

The argument that white-collar workers don't need unions has been disproven. Workers are asking for better treatment as their conditions get worse. There are dozens of recent successful strikes and union drives for tech workers to draw from as examples.

In a non-tech organizing campaign that is close to my heart, Starbucks workers across the country have filed for union elections. As of this writing, over two hundred stores have won elections, including the store in my hometown of Hopewell, New Jersey, and the one by my house in Bloomfield. They have faced incredible union-busting tactics, illegal store closures, and firings, but have won about 85 percent of the elections held so far. Amazon workers won an election as an independent union in Staten Island, New York. The worker-led organizing at Starbucks and Amazon specifically, and the broader renewed interest in unions, gives me hope that if we stick together, working people can overcome our political system and live decent lives without having to work three jobs.

We won our election because we were resilient, we kept quiet around bosses, and we took direction. We worked together during bargaining, and we fought back when management retaliated. It wasn't easy, but we did it. We beat two corporations with unlimited resources because we fought for each other. I'm proud of everything we've accomplished, and I'd do it again, knowing the outcome might feel underwhelming compared to the amount of work we put into it.

In the summer of 2021, HCL and Google told us we weren't going back to the office in Bakery Square. We won the right to future remote work in our contract but only if Google lets us keep our permissions after we return to the office. If our new office is more than three miles from Bakery Square, we will be allowed to work four ten-hour days. But it doesn't look like we will ever go back. Two new hires to Catalogs live in Harrisburg and Chicago. Management will not tell us if they know about plans to stay remote permanently, but they have to give us ninety days' notice if we ever have to report to a physical location. They either expect these workers to uproot their lives for a job, which they claimed in bargaining didn't deserve more than $16 an hour, or they are just refusing to tell us that we are never working on-site anywhere again. The job postings just say "remote." All these minor things that make life a little more bearable for us, policies that should have been in place since the beginning, we had to fight for, for over two years. But we have them now. The extra sick days Google promised haven't materialized, but we are still fighting for them. Our contract states we are entitled to any additional sick time required by Google that is provided to non-unionized employees. During the run up to the election, our supervisors repeatedly told us we would get eight additional, dedicated sick days as part of Google's new TVC policies, but HCL's lawyers decided our accrued PTO counted as sick time. When we tried to contact the person in charge of enforcing the new guidelines for TVCs, we got a canned response from a worker at another temp agency, telling us to talk to our manager. There is no oversight for the improvements Google publicized a few years ago.

When HCL refused to give us time off for COVID-related illnesses, we were able to force them to comply with Pittsburgh's city ordinance mandating paid leave for COVID-19 illnesses and time for testing and vaccination through our union.

I love my union and my coworkers, but I'm fried. For me, it's been hard to decelerate. I've been keyed up for three years, ready to fight for myself and my coworkers, and it's bled into my personal life. I became short-tempered and argumentative. It was tough to turn it off.

HCL hired another supervisor for our team of six people. She's my fifth lead analyst in three years. We answer to her, and she reports to Donald.

I started a second full-time job in late 2021. I worked both jobs for almost two months so I could take my bargained-for additional days off. Then I put in my notice at HCL. I am a full-time technical writer and editor now. I'm left alone to do my work, and I meet my deadlines. My days are less stressful. It took me until age forty to find a permanent, full-time job that allows me to save money and support my family. I am very fortunate, and I don't take it for granted.

I hate to leave, mostly because my boss wanted me to quit. I wanted to stay on just to fight them, but that didn't seem healthy. I did a lot. I worry that our union is losing so many vocal leaders who would take the fight to management. Will was a pain sometimes, but he was never afraid to tell our bosses they were wrong. He would put himself in harm's way to help his coworkers, and I admire that about him. Some of the most outspoken of us have left for more money and less hassle from management. But the union is in good hands and with a good foundation in our first contract.

———————

Quarantine was tough for Gracie too. She was regaining her confidence in school when everyone was sent home. She struggled. After another long wait, I got her a new therapist. She was diagnosed with PTSD resulting from her mother's death and all the memories from their time together that had started coming back in a different context. Her grades were mediocre, and she spent too much time in bed. Everything got so much harder, and I didn't know how to help her maintain her studies or her mental health. Nothing worked. She missed her friends, and I worried she was getting depressed. I tried to get her all the help and support I could, but I constantly felt at a loss.

We did spend more time together. We got on each other's nerves some, but there were more good moments than bad. Early in quarantine, we beat Mario Bros. after a long session that ended past midnight one Friday. She played softball that spring, and it was a welcome escape.

The guy downstairs moved out, May moved in, and the three of us made a home out of our rowhouse in Bloomfield a few months into the pandemic. If the pandemic ever ended, Gracie would be free to have sleepovers.

In the summer of 2021, Gracie found a recreational fast-pitch softball league across the river in Shaler. The drive to the field runs past Gracie's old preschool and the development where she used to live with her mom and her great-grandma. We stop for Gatorade at the same gas station I used to buy cigarettes at years ago after dropping Gracie off with her mom. Some nights I'm flooded with anxiety and memories of extreme stress: driving to pick her up when she was two and not knowing if her mom would be there with her, or sitting around and waiting for an indeterminate amount of time for them to get back from somewhere. The incidents with her mother at preschool orientation. But there were better memories too: Easter egg hunts in the backyard, Halloween, the occasional birthday dinner when we all got along for Gracie's sake.

Once in a while, Gracie will say, "It's weird driving by my old house." But we mostly just talk about the games.

It took her almost a whole fall season and then the following spring season of striking out until something clicked late in the year. With three games to go in the regular season, against West View, she hit a line drive to center field for her first hit of the year. In her next at bat, she blooped one over the second baseman's head. The next game, she had two more hits and a walk. In the season finale, she went one-for-two with a hit by pitch.

The playoffs were held at a nice complex up in Cranberry. I watched the games from beyond the right field fence, pacing during her at bats. She only got out once in three playoff games. Against the top seed, she had three hits, including a clutch two-out, two-run single in an upset win. Rain pushed the next game back, and her team lost momentum. They lost in the semifinals. She hit .700 over her last six games after only getting on base four times via hit by pitch in the first nine games of the season.

She's playing fast-pitch again this fall, and she's hitting at a steadier pace. The field sits down in a valley in a complex next to a playground and a baseball field. The breeze carries a hint of wet grass and bubble gum, and yellow has started to creep into the leaves on the trees beyond the backstop. The days have started getting shorter again. During warm-ups, Gracie laughs and jokes with her teammates. From my spot beyond the outfield fence, it looks easy and natural, and I'm as grateful for that as I am for her doubles down the line.

I don't know how many more nights Gracie and me will have like this. She'll be driving soon. We won't always have the trips to and from games, the discussions about hitting, and the stops at Wendy's. But we still have some time.

Gracie waves at me as she jogs out to her position in left field. I sit in my lawn chair, next to May, and I can hear the ping off the aluminum bats and the chatter from the benches. The players picking each other up and shouting encouragement. Early in the game, there's a lazy fly ball to shallow left, drifting toward the foul line. I stand and watch as Gracie calls for the ball and runs after it.

REFERENCES

late nineties at Microsoft: Ben Tarnoff, *The Making of the Tech Worker Movement* (California: Logic Magazine, 2020).

history of the development: Bakery Square Development (Former Nabisco Factory), Western Pennsylvania Brownfields Center, https://www.cmu.edu/steinbrenner/brownfields/Case%20Studies/pdf/Bakery%20Square.pdf.

what had been a cheap city: Nick Coles, "Black Homes Matter: The Fate of Affordable Housing in Pittsburgh." *Working-Class Perspectives,* February 8, 2016, https://workingclassstudies.wordpress.com/2016/02/08/black-homes-matter-the-fate-of-affordable-housing-in-pittsburgh/.

where the condos and orange houses now sat: Bakery Square 2.0 Development (Former Reizenstein School) Western Pennsylvania Brownfields Center, https://www.cmu.edu/steinbrenner/brownfields/Case%20Studies/pdf/Bakery-Square-2.0-case-study-2013.pdf.

Penn Plaza apartments once stood: Margaret J. Krauss, "News of Penn Plaza's Redevelopment Shook Pittsburgh. What Did the City Know before It Happened?" *90.5 WESA*, November 23, 2020, https://www.wesa.fm/development-transportation/2020-11-23/news-of-penn-plazas-redevelopment-shook-pittsburgh-what-did-the-city-know-before-it-happened.

horror stories from other TVCs about sexual harassment, health care, and the feeling of being voiceless: Mark Bergen and Josh Eidelson, "Inside Google's Shadow Workforce," *Bloomberg News,* July 25, 2018, https://www.bloomberg.com/news/articles/2018-07-25/inside-google-s-shadow-workforce#xj4y7vzkg.

Google also hired an anti-union law firm and started a campaign called Project Vivian: Lauren Kaori Gurley, "Google Had Secret Project to 'Convince' Employees 'that Unions Suck,'" *Vice*, January 10, 2022, https://www.vice.com/en/article/v7d7j9/google-had-secret-project-to-convince-employees-that-unions-suck.

fired, and some quit under duress: Opheli Garcia Lawler, "The Women behind the Google Walkout Say They're Facing Retaliation." *The Cut,* April 22, 2019, https://www.thecut.com/2019/04/women-behind-google-walkout-say-theyre-facing-retaliation.html.

The year after the walkout: Shirin Ghaffary, "It's Been a Year since 20,000 Google Employees Walked Off the Job. And They're Madder than Ever," *Vox,* November 1, 2019, https://www.vox.com/recode/2019/11/1/20942234/google-walkout-one-year-anniversary-unionization-organizing-tech-activism-we-wont-built-it.

Uber spent over $200 million: Edward Ongweso Jr., "Proposition 22's Victory Shows How Uber and Lyft Break Democracy," *Vice,* November 5, 2020, https://www.vice.com/en/article/akddx8/proposition-22s-victory-shows-how-uber-and-lyft-break-democracy.

CWI is even scarier: The Truth about the CWI, https://www.thetruthaboutcwi.com/.

They want to make Proposition 22 national law: Bryce Covert, "Uber Has Shown Us the Future It Wants for Employment," *Slate*, August 16, 2022, https://slate.com/business/2022/08/uber-lyft-gig-economy-worker-flexibility-choice-act.html.

Meanwhile, the national cost of living had increased more than 8 percent: https://www.bls.gov/data/inflation_calculator.htm.

"Management Responds": Lawrence Mishel, Lynn Rhinehart, and Lane Windham, "Explaining the Erosion of Private-Sector Unions." *Economic Policy Institute*, November 18, 2020, https://www.epi.org/unequalpower/publications/private-sector-unions-corporate-legal-erosion/; Robert H. Zieger and Gilbert J. Gall, *American Workers, American Unions*, 3rd ed. (Baltimore: Johns Hopkins University Press, 2002); Howard Zinn, *A People's History of the United States* (New York: Harper Perennial Modern Classics, 2015).

Legislation like the PRO Act: Nick Niedzwiadek and Eleanor Mueller with help from Alex Daugherty, "Unions' Post-Reconciliation PRO Act Push," *Politico*, August 15, 2022, https://www.politico.com/newsletters/weekly-shift/2022/08/15/unions-post-reconciliation-pro-act-push-00051726.

union popularity is on the rise: Justin McCarthy, "U.S. Approval of Labor Unions at Highest Point since 1965," *Gallup,* August 30, 2022, https://news.gallup.com/poll/398303/approval-labor-unions-highest-point-1965.aspx.

One of my favorite organizing strategies: Sam Luebke and Jennifer Luff, "Organizing: A Secret History" *Labor History* 2003 reprinted in *The Forge: Organizing Strategy and Practice,* September 27, 2019, https://forgeorganizing.org/article/organizing-secret-history?page=1.

Starbucks workers across the country: Union Election Data, September 21, 2022, https://unionelections.org/data/starbucks/.

In addition to the works listed above, I read the following books while writing *Team Building*, which greatly informed my work and helped put it into context.

David Graeber, *Bullshit Jobs*: *A Theory* (New York: Simon and Schuster, 2018).

Steven Greenhouse, *Beaten Down, Worked Up: The Past, Present, and Future of American Labor* (New York: Knopf, 2019).

Sarah Jaffe, *Work Won't Love You Back*: *How Devotion to Our Jobs Keeps Us Exploited, Exhausted, and Alone* (New York: Bold Type Books, 2021).

Kim Kelly, *Fight Like Hell*: *The Untold History of American Labor* (New York: Atria/One Signal Publishers, 2022).

Rob Larson, *Bit Tyrants*: *The Political Economy of Silicon Valley* (Chicago: Haymarket Books, 2020).

Jane F. McAlevey, *No Shortcuts: Organizing for Power in the New Gilded Age* (Oxford: Oxford University Press, 2016).

Gabriel Winant, *The Next Shift: The Fall of Industry and the Rise of Health Care in Rust Belt America* (Cambridge: Harvard University Press, 2021).

ACKNOWLEDGMENTS

Thank you to Anne Trubek at Belt Publishing for your endless patience, advocacy, and support. Thanks to Michael Jauchen and Martha Bayne for the thoughtful edits.

Thank you to Kelly Andrews and Taylor Grieshober for your copy edits, encouragement, and feedback on various drafts of the book. Thank you to Derek Green and Rich Gegick for input and advice that helped shape the book and keep me sane. I am lucky to have so many great writers in my corner.

Special thanks to Erin Gatz at Prototype for trusting me with the door code.

Thank you to all my former coworkers in USW Local 4040. We accomplished an amazing feat. I hope I did the experience justice. Special thanks to everyone on the CAT and the Organizing and Bargaining Committees for putting in so much extra work. Thank you to George for getting us through the dicey early days. I miss you all. Catalogs forever.

Thank you to Auni, Todd, Chris, and all the FTEs at Google who supported our efforts, especially the inaugural Alphabet Workers Union leadership team and the organizers and participants of the walkout. Your support was invaluable.

Thank you to everyone at USW: Maria Somma, Mariana Padias, Nate Kilbert, Rachael Davis, Colleen Wooten, Caroline Brickman, Lindsay Dissler, Jeff Cech, Jon Walker, Robin Sowards, Susan Schwartz, and anyone I may have missed. I learned so much from all of you.

Damon Di Cicco, thank you for convincing the higher-ups to take a chance on us and for introducing me to tabletop RPGs. I will see you soon.

Thanks to Dave DeSario at TWJ for the chance to keep a hand in organizing post-HCL; Emma Kinema and Wes McEnany at CWA for all the extra training; and Clarissa Redwine for your support and amazing Kickstarter content. I am deeply indebted to your kindness and solidarity.

Thank you to Mom and Dad for always honking in support of striking workers on picket lines and teaching me the importance of unions. I wish I had half your work ethic and a quarter of your integrity.

Finally, thank you to Amy, Ray, and Cowboy for always being there for me. I love you.

CPSIA information can be obtained
at www.ICGtesting.com
Printed in the USA
JSHW022020260423
40730JS00003B/3

9 781953 368331